Group III-Nitride Semiconductor Optoelectronics

Group III-Nitride Semiconductor Optoelectronics

C. Jayant Praharaj
Band Photonics Materials
CA, US

WILEY

This edition first published 2024

Registered Office
John Wiley & Sons, Inc., 111 River Street, Hoboken, NJ 07030, USA

For details of our global editorial offices, customer services, and more information about Wiley products visit us at www.wiley.com.

Wiley also publishes its books in a variety of electronic formats and by print-on-demand. Some content that appears in standard print versions of this book may not be available in other formats.

A catalogue record for this book is available from the Library of Congress

Hardback ISBN: 9781119708636; ePub ISBN: 9781119708599; ePDF ISBN: 9781119708629;
oBook ISBN: 9781119708681

Cover Image: © zf L/Getty Images
Cover Design: Wiley

Set in 9.5/12.5pt STIXTwoText by Integra Software Services Pvt. Ltd, Pondicherry, India

Contents

Preface

The field of group III-nitride optoelectronic devices is about three decades old. Pioneering efforts by S. Nakamura led to early demonstrations of blue light-emitting diodes and lasers. However, improvements in quantum efficiency, power efficiency, and threshold currents have driven a lot of new innovations spanning these decades. The major technological challenges for optoelectronic emitters in this material system comes in two varieties. The first is the existence of large spontaneous and piezoelectric polarizations and the resulting large built-in fields in quantum wells, quantum wires, and quantum dots. The second is the large acceptor ionization energies and low ionization ratios of acceptors that make getting good *p*-type layers for device optimization difficult.

The first of these challenges can be tackled using device tricks or by using different types of material optimizations like switching to non-polar or semi-polar growth planes. Each way of improving device performance comes with its pros and cons, and the best solutions may arise from a combination of material (e.g. growth, crystallographic) and device methods. This book tries to offer a perspective on these optimization methods while emphasizing the device aspects.

The InGaN (indium gallium nitride) alloy system spans the range of bandgaps from 0.7 eV to 3.43 eV, which makes it attractive for designing solar cells since they can absorb the entire solar spectrum efficiently, all of them being direct bandgap material. Suffice it to say that while the bandgap aspect is attractive, device design challenges and other challenges, like optimization of *p*-type layers, remain. Therefore, most reported experimental work on InGaN photovoltaics has reported modest efficiencies. Most of the serious experimental work on InGaN photovoltaics has happened in the last decade, and a decade is less than enough time to reach efficiencies to compete with or challenge existing photovoltaics. We hope that the chapter on photovoltaics (Chapter 8) serves as a good guide for future research to improve quantum efficiencies and power conversion efficiencies in InGaN solar cells.

This book tries to cover the effect of spontaneous and piezoelectric polarization on quantum wells, quantum wires, and quantum dots with a view to understand the overall effect on device performance. We also hope that the reader can get enough understanding of the device-related polarization phenomena to be able to do device optimization in this material system.

C. Jayant Praharaj
Sunnyvale, 2022

1

Introduction

The group III-Nitride semiconductors (InN [indium nitride], GaN [gallium nitride], and AlN [aluminum nitride]) have achieved a high degree of maturity over the last two or three decades, although specific areas of research still uncover new grounds. The bandgaps of these semiconductors in their wurtzite crystal structure are 0.7 eV for InN, 3.4 eV for GaN, and 6.2 eV for AlN. This means that alloying these semiconductors with one another achieves bandgaps that range from the infrared to the ultraviolet. Further, all three of these are direct bandgap semiconductors (the valence band maxima and conduction band minimum are at the same point in the Brillouin zone). This combination of a wide range of bandgaps and the direct nature of the bandgaps has tremendous implications for optoelectronic devices. The photon or light emission can have high efficiency over a large frequency range as a result. Also, the photon or light absorption can be relatively strong over a wide range of frequencies or wavelengths. The basic conversion factor for this is provided by the Planck relation of quantum mechanics:

$$E_g = h\nu_g$$

where E_g is the bandgap, h is Planck's constant and ν_g is the frequency of absorption or radiation corresponding to the bandgap. It must be kept in mind that frequencies and energies above the bandgap can potentially be emitted or absorbed. However, as explained in Chapter 2, statistical physics involving Fermi–Dirac distributions and density-of-state considerations often mean that a small range above the bandgap is involved in devices like emitters. Photovoltaic devices use absorption over a wide range starting from the bandgap energy.

Although we focus on optoelectronic devices in this book, it should be kept in mind, for perspective, that the group III-Nitride semiconductors have other wide applications like high-power, high-frequency microwave devices. In fact, a lot of the material aspects of group III-Nitride semiconductors were understood and improved upon during research efforts in both microwave devices and optoelectronic devices. High electron mobility transistors and heterojunction bipolar transistors are only some of the devices where high-power microwave operation is possible, for example, in ranges beyond the capability of the arsenides and phosphides. Furthermore, the group III nitrides can potentially be used in the terahertz range, although more research is required. A lot of the same fundamental factors influence physical properties. For example, the impact ionization coefficient involves collisions between charge carriers that excite an electron from the valence band to the conduction band in a three-particle setting, and the larger bandgap leads to lower impact ionization coefficients and larger breakdown voltages. This is the basis for the realization of high-power microwave operation. The same bandgap determines the lower cut-off energy for the absorption coefficient in some optoelectronic devices.

Group III-Nitride Semiconductor Optoelectronics, First Edition. C. Jayant Praharaj.

In Chapter 2, we discuss the basic physical properties of group III-Nitride semiconductors like bandgaps, effective masses, absorption coefficients, polarization, and a host of other properties. We focus mainly on bulk properties because they serve as a good point of departure for more advanced concepts discussed in this book. In Chapter 3, we discuss epitaxial growth of group III nitrides using molecular beam epitaxy and metal organic chemical vapor deposition. Since this is a book on device aspects, this chapter can cover only the basic aspects of these growth techniques, and the reader should refer to papers and books on crystal growth for a deeper perspective. Chapter 4 discusses the very important topic of how the properties are determined by confinement in the form of quantum wells, quantum wires, and quantum dots. It forms the basis of later chapters on light emitters, for example, because a vast majority of them use quantum wells, wires, and dots. Chapter 5 discusses light-emitting diodes and lasers built using group III-Nitride semiconductors. Device parameters like quantum efficiencies and threshold current densities are discussed as they pertain to this specific material system. Chapter 6 discusses inter-sub-band light emitters in this material system. Chapter 7 covers photodetectors in this material system as they cover specific frequency or wavelength ranges. Chapter 8 covers the results so far obtained in the field of photovoltaic materials in the group III-Nitride material system. Factors constraining performance and routes to future improvements are discussed. Chapter 9 covers quantum emitters using group III-Nitride materials. Chapters 10 to 12 cover some recent topics like strong-coupling polaritons, plasmonic devices and photonic integrated circuits.

Group III-Nitride optoelectronic devices have a wide range of applications like optical data storage, solid state lighting, light detection and ranging (LiDAR), and scientific metrology. As the device parameters are better understood and as device improvements take place, the range of applications expands both in terms of type and capability. The following chapters cover the most basic physics and the device aspects needed to understand current devices and potential future applications.

2

Band Structure and Optical Properties of Group III-Nitride Semiconductors

Crystal Symmetry (Wurtzite and Cubic Phases)

Wurtzite Group III-Nitride Semiconductors

Wurtzite semiconductors have hexagonal crystal symmetry with distorted tetrahedral bonding. The bond length along the *c*-axis, which is perpendicular to the hexagonal planes, is longer than the bond lengths at an angle to the *c*-axis. The near tetrahedral bonding repeats itself along the *c*-axis as shown in Figure 2.1.

Cubic Group III-Nitride Semiconductors

While the wurtzite crystal structure is the most thermodynamically stable for indium nitride (InN), gallium nitride (GaN), and aluminum nitride (AlN), the cubic phase can also be grown under special circumstances. The crystal structure of the cubic phases is the zinc-blend structure found in, for example, GaAs (gallium arsenide) and InAs (indium arsenide). In the zinc-blend structure, each group III atom (for example, Ga) forms four tetrahedral bonds with four group V atoms (for example, N). The crystal structure in this case consists of two interspersed, face-centered cubic lattices.

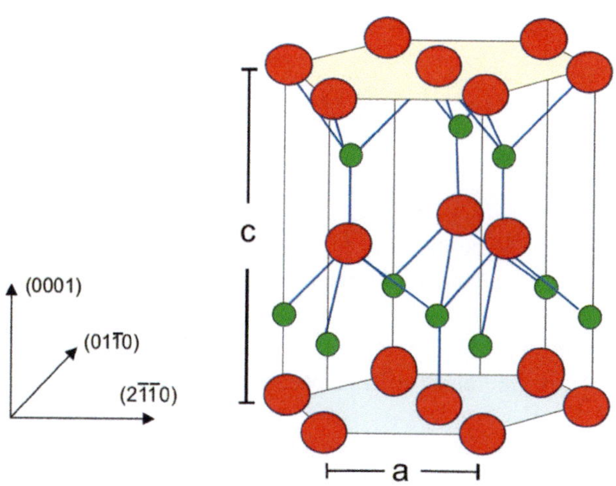

Figure 2.1 Wurtzite crystal structure of group III nitrides. The large atom is the metallic atom (In, Ga, or Al), and the smaller atom is nitrogen.

Group III-Nitride Semiconductor Optoelectronics, First Edition. C. Jayant Praharaj.

Lattice Periodicity and Crystal Hamiltonian

Periodic Crystal Symmetry Effect on Electron States

The effective quantum mechanical Hamiltonian of the semiconductor crystal, which determines the allowed electron energy states, reflects the periodicity of the crystal lattice and is given by the following form

$$H = -\frac{\hbar^2}{2\,m_0}\nabla^2 + V_0(\boldsymbol{r}) \tag{2.1}$$

where ∇^2 is the Laplacian operator and the first term is the kinetic energy operator. The symbol $\hbar$ is the reduced Planck constant. V_0 is the periodic crystal potential. The periodicity of the crystal can be expressed mathematically by means of the translational symmetry condition

$$V_0(\boldsymbol{r} + \boldsymbol{R_n}) = V_0(\boldsymbol{r}) \tag{2.2}$$

where $\boldsymbol{R_n}$ is a translation vector joining two lattice points at which the ions are located.

In reality, since the crystal can contain on the order of 10^{23} atoms and electrons, the actual Hamiltonian is a many-body Hamiltonian, which has in it the kinetic energies of the electrons, the interaction of each electron with each ion, and the Coulomb interaction of each electron with every other electron. The effective single electron Hamiltonian viewed above is a simplification of the above many-body problem and its justification is in good agreement with experiments.

The allowed single-particle energy states are given by the eigenvalues of the time-independent Schrodinger equation

$$H_\Psi = E_\Psi \tag{2.3}$$

In practice, there is an infinite number of solutions for the above eigenvalue-eigenvector equation given by (E_i, Ψ_i) where i ranges from one to infinity.

Bloch's Theorem and Nature of Electron States

Bloch's theorem states that the eigenvectors of the crystal Hamiltonian (with space-periodic potential profile) have the form

$$\psi_{\boldsymbol{k}}(\boldsymbol{r}) = exp(i\boldsymbol{k}.\boldsymbol{r})u_k(\boldsymbol{r}) \tag{2.4}$$

where $u_k(\boldsymbol{r})$ is a spatially periodic function with the periodicity of the crystal lattice. The $\boldsymbol{k}$ vector takes on a range of values allowed by periodic boundary conditions based on crystal size that are imposed on the wave function. While the actual crystal has finite boundaries in the form of surfaces, these periodic boundary conditions give good physical results and are useful for solving most of the electronic and optoelectronic problems, like electron transport in electric fields, optical response functions, etc.

The Bloch states above exist in the form of bands, so that the same $\boldsymbol{k}$ value can give a different wave function and different energy in another band. The band index can be added to write the Bloch function as

$$\psi_{nk}(\boldsymbol{r}) = exp(i\boldsymbol{k}.\boldsymbol{r})u_{\boldsymbol{nk}}(\boldsymbol{r}) \tag{2.5}$$

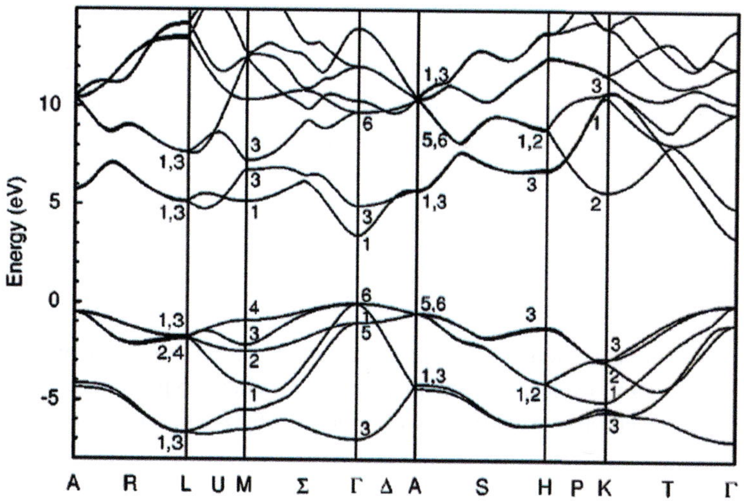

Figure 2.2 *E*-***k*** relations for wurtzite GaN along symmetry directions – from Reference [1], used with permission.

The energy corresponding to the Bloch function $\psi_{n\boldsymbol{k}}(\boldsymbol{r})$ is a function of the $\boldsymbol{k}$ vector, so that the energy dispersion relations of the crystal can be written as

$$E_{\psi_n}(\boldsymbol{k}) = E_n(\boldsymbol{k}) \tag{2.6}$$

The Brillouin Zone and Range of k Values

The wave-vector $\boldsymbol{k}$ has physical meaning within a range of values that constitute the Brillouin zone of the material. The boundary of the Brllouin zone corresponds to reciprocal lattice vectors. Bloch wave functions with $\boldsymbol{k}$ values that differ by a reciprocal lattice vector can be proven to be identical.

The energy dispersion relation has periodicity properties given by

$$E_n(\boldsymbol{k} + \boldsymbol{G}) = E_n(\boldsymbol{k}) \tag{2.7}$$

where $\boldsymbol{G}$ is a reciprocal lattice vector. Figure 2.2 shows a typical E-$\boldsymbol{k}$ diagram for GaN.

Quantum Mechanical Properties Corresponding to Bloch States

The wave function gives useful mechanical information like the expected value of velocity and momentum. For example, the expected value of velocity, which is given by the integral

$$v(\boldsymbol{k}) = \int \Psi^*_{nk}\left(-\frac{i\hbar\nabla}{m}\right)\Psi_{nk} d^3r \tag{2.8}$$

can be proven to have the value

$$v(\boldsymbol{k}) = \frac{1}{\hbar}\nabla_{\boldsymbol{k}} E(\boldsymbol{k}) \tag{2.9}$$

Valence Band and Conduction Band

In semiconductors, the highest occupied band at a temperature of absolute zero is fully occupied and is called the valence band. A bandgap separates it from the lowest unoccupied band called the conduction band. At finite temperature, the occupancy of electron states is given by the Fermi–Dirac statistical distribution given by

$$f\left(E_n(\mathbf{k})\right)=\frac{1}{exp\left(\dfrac{E_n(\mathbf{k})-E_F}{kT}\right)+1} \tag{2.10}$$

where E_F is the Fermi level or the chemical potential of the semiconductor. The value of f is between 0 and 1 and is consistent with the Pauli exclusion principle. Figure 2.3 shows typical bandgaps versus lattice constants of nitride semiconductors as well as some other common semiconductors.

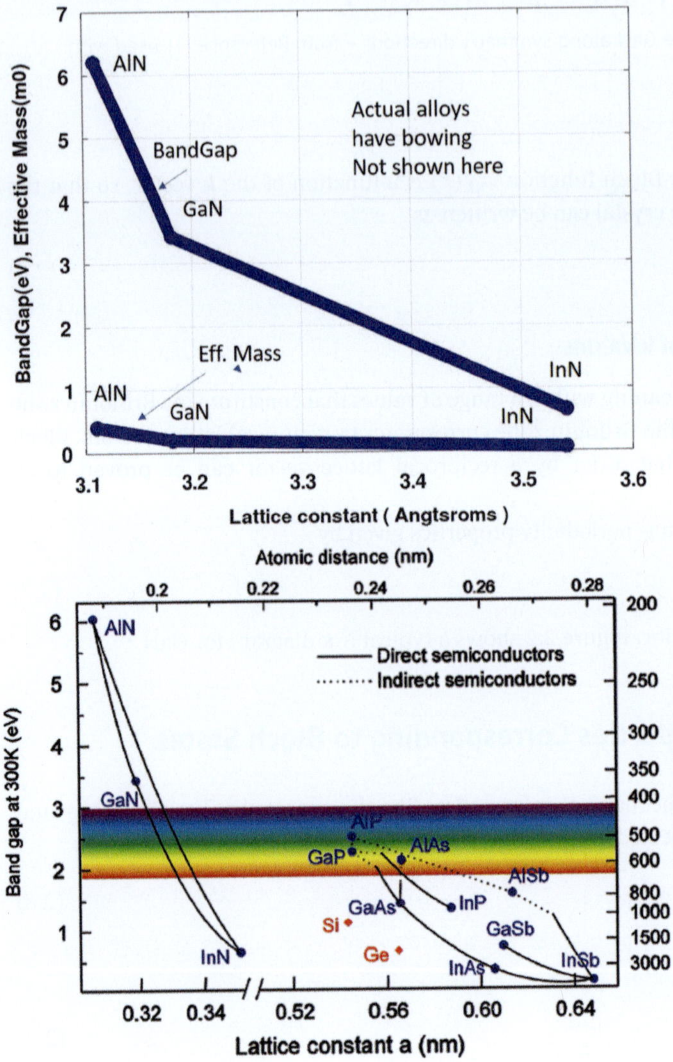

Figure 2.3 Bandgaps, effective masses, and wavelengths corresponding to bandgap for group III nitrides and some comparisons with other semiconductors.

A semiconductor can have electrons in the conduction band at finite temperature due to the Fermi–Dirac distribution function. A semiconductor without any impurities added is called an intrinsic semiconductor. When impurities are added, donor energy levels are introduced close to the conduction band and acceptor energy levels are introduced close to the valence band. When a valence band level is unoccupied with electrons, it is called a hole. It can be proven that the electrical current due to the electron velocities is zero in a fully filled band. The hole picture makes it easier to describe the current conduction in the valence band in terms of moving effective positive charges.

Silicon acts as a donor in group III-Nitride semiconductors by substituting for the metallic atom (In, Ga, or Al). Acceptors are typically obtained in group III-Nitride semiconductors using group II metals like Mg, Ca, and Sr. These typically substitute the metallic atom and give holes in the valence band when used as dopants.

Light–Matter Interaction in Semiconductors

Perturbation Due to Electromagnetic Field

In the presence of light, the electromagnetic field modifies the Hamiltonian of the total system. The semiclassical light–matter Hamiltonian can be derived to be of the form

$$H=\frac{(\boldsymbol{p}-e\boldsymbol{A})^2}{2m}+e\varphi+V_0(\boldsymbol{r}) \tag{2.11}$$

where the quantum mechanical momentum operator is given by

$$\boldsymbol{p}=-i\hbar\nabla \tag{2.12}$$

where $\boldsymbol{A}$ is the vector potential for the electromagnetic field, φ is the scalar potential, and V_0 is the usual periodic crystal potential. Since a gauge can be chosen for the electromagnetic field, the potential term can be set to zero and the Hamiltonian becomes

$$H=\frac{(\boldsymbol{p}-e\boldsymbol{A})^2}{2m}+V_0(\boldsymbol{r}) \tag{2.13}$$

where the electric field is given by

$$\boldsymbol{E}=-\frac{\partial\boldsymbol{A}}{\partial t} \tag{2.14}$$

For plane electromagnetic (EM) waves, the vector potential is given by

$$A=A_0exp(i(\boldsymbol{k}.\boldsymbol{r}-\omega t)) \tag{2.15}$$

$$\boldsymbol{E}=i\omega\boldsymbol{A} \tag{2.16}$$

The total Hamiltonian can then be written as an unperturbed part plus a perturbation Hamiltonian

$$H=H_0+H' \tag{2.17}$$

where H' is given by

$$H'=-e\frac{\boldsymbol{p}.\boldsymbol{A}+\boldsymbol{A}.\boldsymbol{p}}{2m}+\frac{e^2\boldsymbol{A}^2}{2m} \tag{2.18}$$

A further gauge choice called the Coulomb gauge can be made, and the perturbation Hamiltonian can be written as

$$H' = -\frac{\boldsymbol{p}.\boldsymbol{A}}{m} + \frac{e^2\boldsymbol{A}^2}{2m} \tag{2.19}$$

The $\boldsymbol{A}^2$ term has a much weaker effect compared to the first term so that most aspects of light–matter interaction can be understood in terms of the first term of this perturbation Hamiltonian. The $\boldsymbol{p}.\boldsymbol{A}$ term gives rise to single-photon emission and absorption processes while the $\boldsymbol{A}^2$ term represents double-photon processes.

Transitions Due to the Electromagnetic Field

The time-dependent Schrodinger equation

$$i\hbar\frac{\partial\psi}{\partial t} = H\Psi \tag{2.20}$$

needs to be solved to obtain the response of the semiconductor in the presence of EM fields. Time-dependent perturbation theory provides the transition rate from one state to another state due to the electromagnetic field. It can be written in terms of the Fermi's golden rule

$$R(i,f) = \frac{2\Pi}{\hbar}\left|\langle\psi_f|H'|\psi_i\rangle\right|^2 \delta(E_f - E_i \mp \hbar\omega) \tag{2.21}$$

Here, the delta function ensures conservation of energy between the matter system and the light system. Here, ħω is the energy of the photon quantum of the electromagnetic field and the plus and minus sign refer to the absorption and emission of light respectively. $\langle\psi_f\ |H'|\ \psi_i\rangle$ is a notation for the matrix element of the perturbation Hamiltonian between initial and final states of the semiconductor.

$$\langle\Psi_f\ |H'|\ \Psi_i\rangle = \int \Psi_f(\boldsymbol{r})H'\Psi_i(\boldsymbol{r})d^3\boldsymbol{r} \tag{2.22}$$

Further, the total transition rate from the initial state can be given by integrating over the density of states (DOS). The absorption rate and the stimulated emission rate are given by

$$R_{abs} = \frac{2\Pi}{\hbar}\int \left|\langle\Psi_c|H'|\Psi_v\rangle\right|^2 \delta(E_c - E_v - \hbar\omega)g(E_c)dE_c \tag{2.23}$$

$$R_{ems} = \frac{2\Pi}{\hbar}\int \left|\langle\Psi_v|H'|\Psi_c\rangle\right|^2 \delta(E_v - E_c - \hbar\omega)g(E_c)dE_c \tag{2.24}$$

Spontaneous Emission

An electron in an excited state can emit a photon of electromagnetic energy even when no incident radiation is present. This is known as a spontaneous emission. The derivation of spontaneous emission requires that the electromagnetic field be quantized in addition to the electron energy states. The reader is referred to an advanced textbook on quantum mechanics to understand the details of electromagnetic field quantization.

When the electromagnetic field is quantized, the total emission rate is proportional to $(n+1)$ where n is the number of photons in the electromagnetic mode causing the transition. The stimulated emission term is proportional to n while the spontaneous emission term is 1 n^{-1} of the stimulated emission term. Some books derive the spontaneous emission term using thermodynamic arguments involving so-called Einstein A and B coefficients.

k-Vector Conservation in Crystal Transitions

Since the conduction and valence band wave functions are Bloch functions in the crystal, the matrix element become proportional to a delta function in the $\boldsymbol{k}$ vectors

$$\langle \psi_c | H' | \psi_v \rangle = \langle u_{kc} | H' | u_{kv} \rangle \delta(\boldsymbol{k_c} - \boldsymbol{k_v} - \boldsymbol{k_{photon}}) \tag{2.25}$$

Since the wave vector of a photon of energy nearly equal to the bandgap is several times smaller than the Brillouin zone boundary $\boldsymbol{k}$ vector, the delta function reduces to approximately $\delta(\boldsymbol{k_c} - \boldsymbol{k_v})$.

$$\langle \psi_c | H' | \psi_v \rangle \approx \langle u_{kc} | H' | u_{kv} \rangle \delta(\boldsymbol{k_c} - \boldsymbol{k_v}) \tag{2.26}$$

The $\boldsymbol{k}$-vector conservation has implications for light emission and light absorption by semiconductors. The $\boldsymbol{k}$-vector conservation requirement means that transitions must occur vertically in $\boldsymbol{k}$ space. That means direct bandgap semiconductors, where the minimum of the conduction band and maximum of the valence band are at the same $\boldsymbol{k}$ vector in the Brillouin zone, will be efficient light emitters when states close to the bottom of the conduction band are populated with electrons and states close to the top of the valence band are populated with holes. The group III-Nitride semiconductors InN, GaN, and AlN are all direct bandgap semiconductors. This is true for both the hexagonal wurtzite III nitrides and for cubic III nitrides. Therefore, the group III-Nitride semiconductors are useful for optoelectronic applications provided other device issues and materials issues are addressed.

Absorption Coefficient and Optical Gain

The net transition rate per unit volume from the valence band to the conduction band can then be written as a sum over all initial valence band states and all final conduction band states

$$R_{v \to c} = \frac{2}{V} \sum_{kc} \sum_{kv} \frac{2\Pi}{\hbar} |H'_{cv}|^2 \delta(E_c - E_v - \hbar\omega) f_v (1 - f_c) \tag{2.27}$$

where f_v and f_c are the Fermi–Dirac occupation factors, which take into account the fact that the initial valence band state must be occupied and the final conduction band state must be empty due to the Pauli exclusion principle. Similarly, the total transition rate from the valence band to the conduction band can be written as

$$R_{c \to v} = \frac{2}{V} \sum_{\boldsymbol{kc}} \sum_{\boldsymbol{kv}} \frac{2\Pi}{\hbar} |H'_{cv}|^2 \delta(E_c - E_v - \hbar\omega) f_c (1 - f_v) \tag{2.28}$$

The net transition rate per unit volume is

$$R_{v \to c} - R_{c \to v} = \frac{2}{V} \sum_{\boldsymbol{kc}} \sum_{\boldsymbol{kv}} \frac{2\Pi}{\hbar} |H'_{cv}|^2 \delta(E_c - E_v - \hbar\omega)(f_v - f_c) \tag{2.29}$$

The presence of scattering from the final state during the transition results in the energy-conserving delta function becoming a Lorentzian line-shape function

$$\delta\left(E_c - E_v - \hbar\omega\right) \to \frac{\frac{\Gamma}{2\Pi}}{\left(E_c - E_v - \hbar\omega\right)^2 + \left(\frac{\Gamma}{2}\right)^2} \tag{2.30}$$

Complex Dielectric Function

The full optical response of the material to any incident electromagnetic radiation is characterized by a complex dielectric function or complex permittivity function

$$\varepsilon(\omega) = \varepsilon_1(\omega) + i\varepsilon_2(\omega) \tag{2.31}$$

The absorption coefficient can be expressed in terms of the imaginary part of the dielectric function

$$\begin{aligned} \alpha &= 2\mathrm{Im}\ \omega\sqrt{\mu\left(\varepsilon_1 + i\varepsilon_2\right)} \\ &\approx 2\mathrm{Im}\ \omega\frac{n_r}{c}\left(1 + \frac{i\varepsilon_2}{2\varepsilon_1}\right) \\ &= \frac{\omega}{n_r c}\frac{\varepsilon_2}{\varepsilon_0} \end{aligned} \tag{2.32}$$

when $|\varepsilon_2| \ll \varepsilon_1$. The real part of the dielectric function can be found using the Kramers–Kronig relation

$$\varepsilon_1\left(\omega\right) = \varepsilon_1\left(\omega\right) + \frac{2}{\Pi}P\int_0^{\infty}\frac{\omega'\varepsilon_2\left(\omega'\right)}{\omega'^2 - \omega^2}d\omega' \tag{2.33}$$

where P denotes the principal value of the integral.

The real part of the dielectric function decides the dispersion in the semiconductor while positive and negative values of the imaginary part of the dielectric function decide the absorption or gain in the material.

We use the notation $\boldsymbol{p}_{cv}$ to denote the momentum matrix element between the cell-periodic parts of the conduction and valence band wave functions. With this notation, the absorption coefficient can be calculated as

$$\alpha\left(\hbar\omega\right) = \frac{\pi e^2}{n_r c\varepsilon_0 m_0^2\omega}\frac{2}{V}\sum_k \left|\boldsymbol{e}.\boldsymbol{p}_{cv}\right|^2 \delta\left(E_c - E_v - \hbar\omega\right)\left(f_v - f_c\right) \tag{2.34}$$

where E_c and E_v correspond to the same $\boldsymbol{k}$ vector, as imposed by the $\boldsymbol{k}$-vector delta function after one summation.

The $\boldsymbol{k}$ summation can be carried out using the $\boldsymbol{k}$-space DOS, which is given by

$$g\left(\boldsymbol{k}\right) = \frac{V}{\left(2\pi\right)^3} \tag{2.35}$$

since the periodic boundary conditions restrict the $\boldsymbol{k_x}$, $\boldsymbol{k_y}$, and $\boldsymbol{k_z}$ values to integral multiples of $2\frac{\pi}{L}$. With this, for a semiconductor close to $T = 0\,\mathrm{K}$, the absorption coefficient is reduced to

$$\alpha\left(\hbar\omega\right) = \frac{\pi e^2}{n_r c\varepsilon_0 m_0^2\omega}\left|\boldsymbol{e}.\boldsymbol{p}_{cv}\right|^2 \rho_r\left(\hbar\omega - E_g\right) \tag{2.36}$$

where ρ_r is the reduced or joint DOS

$$\rho_r\left(\hbar\omega - E_g\right) = \frac{1}{\left(2\pi\right)^2}\left(\frac{2m_r}{\hbar^2}\right)^{3/2}\sqrt{\hbar\omega - E_g} \tag{2.37}$$

For many intrinsic semiconductors, the absorption coefficient at $T = 300\,\text{K}$ does not differ very much from the $T = 0\,\text{K}$ absorption coefficient because large-enough bandgaps mean that the valence band is mostly full and the conduction band is mostly empty.

When a semiconductor is optically pumped, or in a device where electrical current injection can lead to non-equilibrium carrier concentrations, the Fermi–Dirac occupation factors are determined by quasi-Fermi levels for conduction and valence band states. Positive gain (or negative absorption coefficient) is achieved when the following inequality is attained

$$f_v < f_c,$$

which is equivalent to

$$E_{\text{fc}} - E_{\text{fv}} > E_c - E_v = \hbar\omega \tag{2.38}$$

This condition is also known as population inversion since in thermal equilibrium, the opposite inequality holds. As we will see in Chapter 5 on group III-Nitride lasers, the population inversion condition, which gives rise to optical gain, can be achieved under non-equilibrium current injection condition.

Spontaneous and Piezoelectric Polarization

The crystal symmetry of the wurtzite group III-Nitride semiconductors (InN, GaN, and AlN) allows the existence of spontaneous polarization in the crystal. The spontaneous polarization points along the *c*-axis. For an infinite crystal, the definition of the spontaneous polarization poses some technical problems having to do with non-uniqueness. However, the modern theory of polarization has been able to define a unique polarization value for wurtzite semiconductors and calculated values are in good agreement with experiments. The reader should keep in mind that the physically significant quantities are polarization discontinuities at interfaces that give rise to bound interface sheet charges. Since the spontaneous polarization value is a function of temperatures, these materials are also called pyroelectric.

$$\begin{aligned} P_{\text{sp}}\text{AlGaN}(x) &= -0.090\text{x} - 0.034(1-x) + 0.021\text{x}(1-x)\text{cm}^{-2} \\ P_{\text{sp}}\text{InGaN}(x) &= -0.042\text{x} - 0.034(1-x) + 0.037\text{x}(1-x)\text{cm}^{-2} \\ P_{\text{sp}}\text{AlInN}(x) &= -0.090\text{x} - 0.042(1-x) + 0.070\text{x}(1-x)\text{cm}^{-2} \end{aligned} \tag{2.39}$$

The cubic phases do not have spontaneous polarization. However, the preferred crystalline form for group III nitrides is the wurtzite, and it is difficult to obtain good device quality cubic materials in this material system. Most devices are, therefore, built from wurtzite layers and spontaneous polarization is very much a design parameter for device optimization. Calculated spontaneous polarization values are shown in Figure 2.4.

Strain introduces additional polarization known as piezoelectric polarization, which is given by non-linear functions that depend on the strain values in the material. The elasticity relations in the material can be expressed as

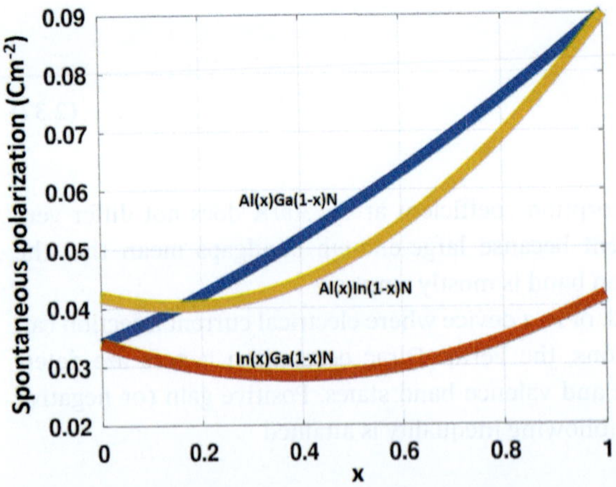

Figure 2.4 Spontaneous polarization for *c*-axis growth of group III-Nitride semiconductors.

$$\sigma_{ij} = \sum\nolimits_{k,l} C_{ijkl}\varepsilon_{kl} \tag{2.40}$$

where σ denotes the stress, ε denotes the strain, and C_{ijkl} are elastic coefficients of the material. Crystal symmetry allows further simplification of the above relation to

$$\sigma_i = \sum\nolimits_{k,l} C_{ij}\varepsilon_j \tag{2.41}$$

$$C_{ij} = \begin{pmatrix} C_{11} & C_{12} & C_{13} & 0 & 0 & 0 \\ C_{12} & C_{11} & C_{13} & 0 & 0 & 0 \\ C_{13} & C_{13} & C_{33} & 0 & 0 & 0 \\ 0 & 0 & 0 & C_{44} & 0 & 0 \\ 0 & 0 & 0 & 0 & C_{44} & 0 \\ 0 & 0 & 0 & 0 & 0 & \frac{1}{2}(C_{11}-C_{12}) \end{pmatrix}$$

The piezoelectric polarizations in group III-Nitride epilayers are given by

$$\begin{aligned}
&P^{pz}_{\mathrm{AlGaN/InN}}(x) = -0.28x - 0.113(1-x) + 0.042x(1-x)\mathrm{Cm}^{-2} \\
&P^{pz}_{\mathrm{AlGaN/GaN}}(x) = -0.0525x + 0.0282x(1-x)\mathrm{Cm}^{-2} \\
&P^{pz}_{\mathrm{AlGaN/AlN}}(x) = 0.026(1-x) + 0.0248x(1-x)\mathrm{Cm}^{-2} \\
&P^{pz}_{\mathrm{InGaN/InN}}(x) = -0.113(1-x) - 0.0276x(1-x)\mathrm{Cm}^{-2} \\
&P^{pz}_{\mathrm{InGaN/GaN}}(x) = 0.148x - 0.0424x(1-x)\mathrm{Cm}^{-2} \\
&P^{pz}_{\mathrm{InGaN/AlN}}(x) = 0.182x + 0.026(1-x) - 0.0456x(1-x)\mathrm{Cm}^{-2} \\
&P^{pz}_{\mathrm{AlInN/InN}}(x) = -0.28x + 0.104x(1-x)\mathrm{Cm}^{-2} \\
&P^{pz}_{\mathrm{AlInN/GaN}}(x) = -0.0525x + 0.148(1-x) + 0.0938x(1-x)\mathrm{Cm}^{-2} \\
&P^{pz}_{\mathrm{AlIn/AlN}}(x) = 0.182(1-x) + 0.092x(1-x)\mathrm{Cm}^{-2}
\end{aligned} \tag{2.42}$$

Figure 2.5 shows calculated piezoelectric polarization values of InGaN on three different substrates.

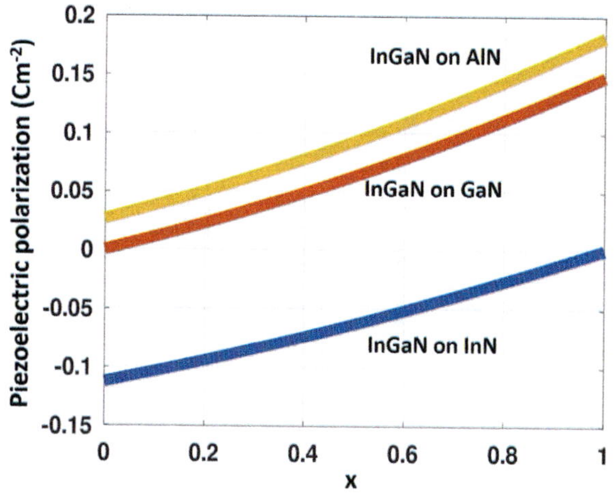

Figure 2.5 Piezoelectric polarization for InGaN on different substrates.

Phonon Spectrum

The phonon spectrum of the group III nitrides shows high values for the optical phonon energies compared to the arsenides and phosphides. AlN has an optical phonon energy of 120 meV, GaN of 90 meV, and InN of 70 meV. For comparison, the optical phonon energy in GaAs is about 30 meV. The range spanned by the acoustic phonons is correspondingly larger compared to the other two material systems. The occupation number for acoustic and optical phonons is given by a Bose–Einstein factor. Figure 2.6 shows phonon dispersion curves in the group III-Nitride semiconductors.

The thermal energy at 300 K is about 25 meV, and therefore, the optical phonon occupation number is small at room temperature. That means absorption of optical phonons is not a significant source of scattering. However, when the device is in non-equilibrium due to applied bias, energetic carriers can emit optical phonons, and this can be a significant source of fast energy relaxation in devices like quantum well light-emitting diodes (LEDs) and lasers. The Boltzmann transport equation or the relevant quantum Boltzmann transport equation determines the carrier-distribution averaged scattering relaxation rates in devices and their effect on device parameters.

Scattering Mechanisms

The charge carriers in the group III-Nitride semiconductors are the electrons in the conduction band and the light holes, heavy holes, and spin-orbit split-off holes in the conduction band. Both electrons and holes undergo scattering whenever there is any departure from the periodic potential. The various sources of these departures are acoustic phonons, optical phonons, donor impurities, acceptor impurities, dislocations, alloy disorder interface roughness, etc.

Since acoustic phonons have a dispersion curve that spans the range from 0 meV to several tens of meVs, they are responsible for electron and hole scattering with absorption as well as emission processes. The most basic type of acoustic phonon scattering is called deformation potential acoustic phonon scattering and is due to modulation of the periodic potential due to the strain field caused by the phonons. Since InN, GaN, and AlN are polar in nature, there is an additional source of acoustic phonon scattering called piezoelectric scattering.

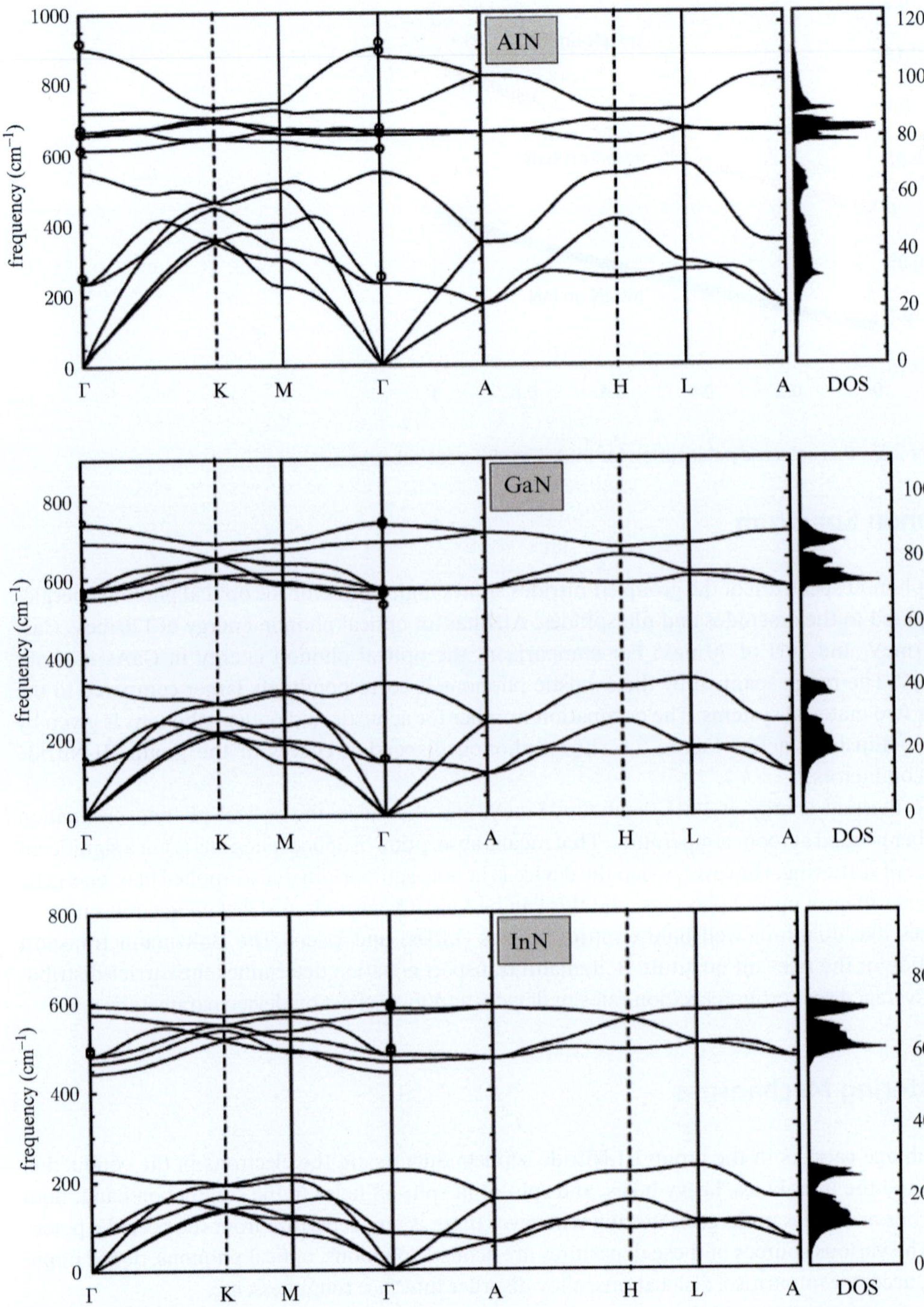

Figure 2.6 Phonon dispersion curves for AlN, GaN, and InN. From Reference [2], used with permission.

The most significant type of optical phonon scattering is called polar optical phonon scattering and is due to the long-range polarization and electric fields caused by optical phonons. In wurtzite materials, the crystal symmetry causes the optical phonon scattering to be due to longitudinal optical (LO)-like and transverse optical (TO)-like scattering. The LO-like scattering rate is dominant. Some theoretical studies have also found contributions to phonon scattering from dynamical quadrupoles in wurtzite GaN. Figures 2.7 and 2.8 show typical phonon scattering rates in gallium nitride.

The cladding layers in LEDs and lasers are doped n-type and p-type on opposite sides. Ionized acceptors and ionized donors cause impurity scattering due to the screened Coulomb potential caused by these ionized dopants. Ionized impurity scattering is a decreasing function of energy.

Since lattice-mismatched buffers and device layers are parts of the overall device structure, dislocations can cause additional scattering. However, good quality devices try to keep dislocation densities below 10^7 per cm^2, and this scattering can often be neglected.

Alloy scattering is an additional source of scattering and is due to the fact that in ternary or quaternary alloys, the actual positions of the metal ions are random. That means there are random fluctuations of the crystal potential about some average, as shown in Figure 2.9.

The scattering rates depend on material parameters like deformation potential, phonon energy, phonon dispersion, alloy disorder pseudopotential, impurity concentration, and DOS, which depend on the effective mass. The vast majority of light emitters are made from quantum wells, quantum wires, and quantum dots (QDs). The DOS is modified in quantum wells, wires, and dots compared to the bulk. This is discussed in more detail in Chapter 4. The modified DOS affects the

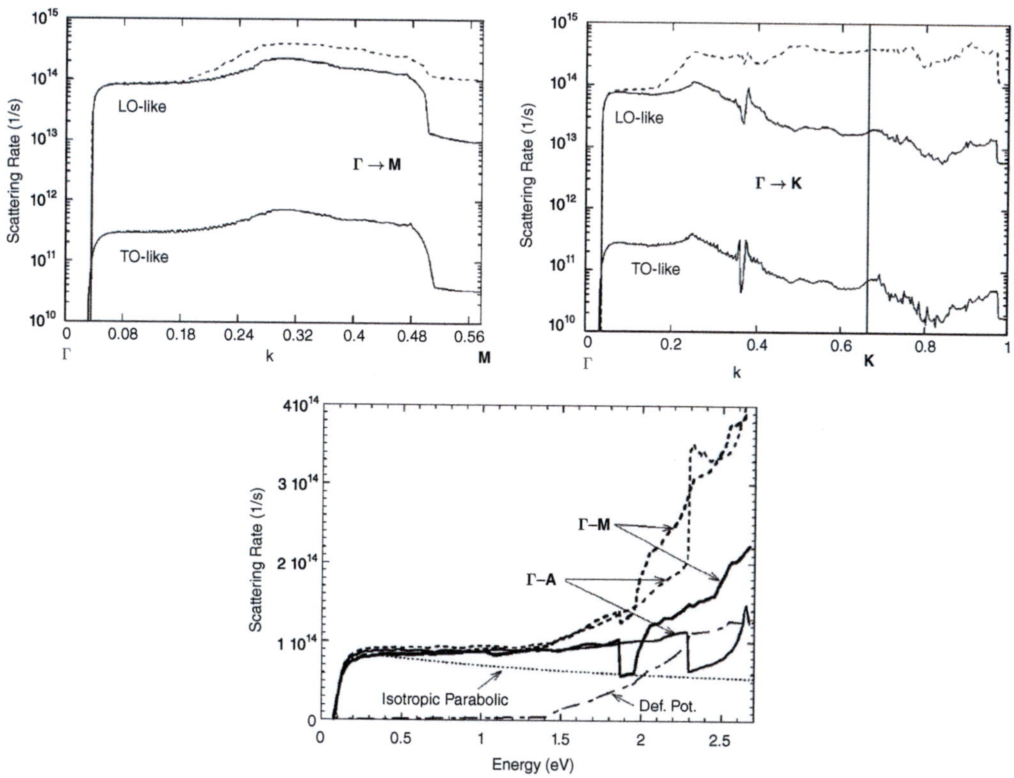

Figure 2.7 (top) Scattering rates from LO-like and TO-like optical phonons in GaN versus energy (bottom). Polar optical phonon plus intervalley scattering comparison with deformation potential scattering in GaN, from Reference [3], used with permission.

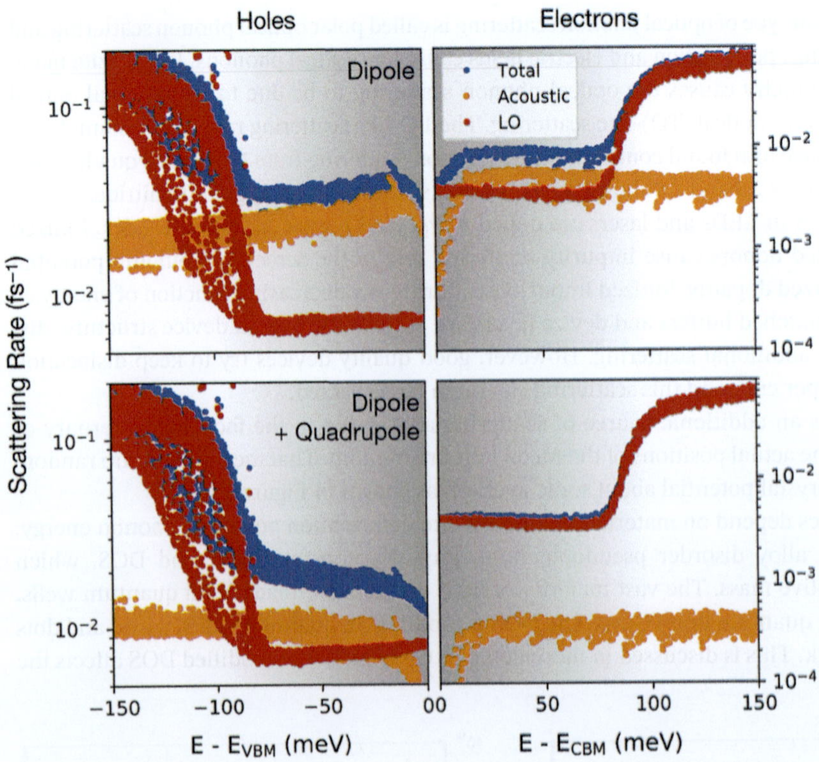

Figure 2.8 Dipolar and quadrupolar scattering rates for electrons and holes in GaN, from Reference [4], used with permission.

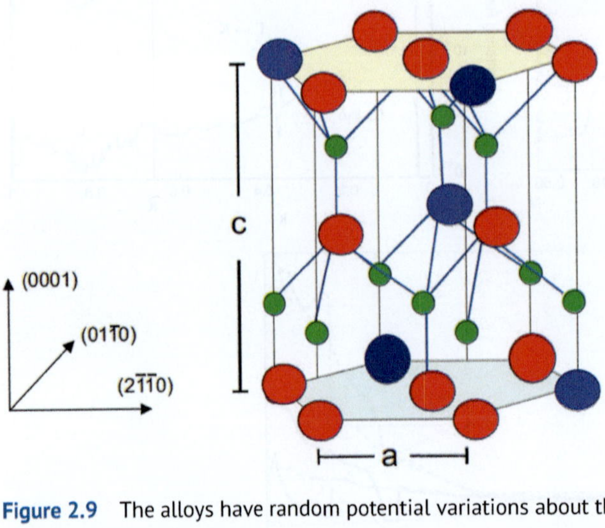

Figure 2.9 The alloys have random potential variations about the average.

scattering rates. Typically, the lower the dimensionality, the lower the DOS and lower the scattering rate. However, careful calculations are always preferable to obtain magnitudes of scattering rates. The reader is referred to References [5–7] for the fundamental physics of bands in solids and scattering processes.

Electron-hole Plasmas and Plasmons

The active layer of LEDs and lasers have large two-dimensional (2D) electron gas (EG) densities and large 2D hole gas (HG) densities under large forward bias, which is needed in typical operation (as explained in Chapter 5). The EG and the HG have plasmon frequencies that depend on the 2DEG density and the 2DHG density, respectively.

When only the first sub-band is occupied, the plasmon frequency is given by

$$\omega_p = \sqrt{\frac{e^2 n_s q}{2\varepsilon_0 m^*}}$$

where q is the wave vector. With the occupation of other sub-bands, the dielectric function becomes a complicated function of frequency and more complicated plasmon frequencies have to be derived. The reader can find more details of quasi-2D and quasi-one-dimensional plasmons in Reference [8]. The plasmons are damped by dissipative mechanisms like phonon scattering. The plasmons are long-lived excitations only when the dissipative rates are much smaller than the plasmon frequencies themselves. When the plasmon frequency is close to the optical phonon frequencies, they hybridize with, or couple to, the phonons and form phonon–plasmon coupled modes. The phonon scattering rates mentioned earlier in the chapter are modified by coupled plasmon–phonon mode scatterings when this happens. Numerical calculations of relaxation rates and Monte Carlo calculations of relaxation rates must take this coupling into account in these cases so that proper comparison with experiments can be done.

Hole plasmons are much more complicated compared to electron plasmons because there are three different types of holes – heavy holes, light holes, and spin-orbit split-off holes, each of which has its own plasmon frequency. In fact, the three different kinds of holes tend to be coupled when the 2D dispersions are calculated. This kind of coupling will be described in more detail in Chapter 4. Since hole effective masses tend to be much greater than electron effective masses, hole plasmons are damped more compared to electron plasmons. However, whether or not hole plasmons play a role in device operation depends on actual device structure and bias conditions.

The phonon–plasmon scattering rates become crucial to determining if energetic electrons and holes injected from the barriers into the well layer in multi-quantum-well devices can relax fast enough so that modulation is decided by other parameters like differential gain (to be defined later in the book). This can become especially important in quantum wire lasers. In QD LEDs and lasers, the scattering from phonons or plasmon phonons is constrained by the phonon-blockade effect (to be discussed in more detail in Chapter 4), where phonon scattering or phonon–plasmon scattering is severely restricted because energy conservation and quasi-momentum conservation cannot be satisfied simultaneously. In quantum wire LEDs and lasers, the phonon–plasmon scattering rates must be calculated very precisely to determine their potential impact on device operation and device speed.

Donors and Deep Acceptors

n-Type and *p*-type doping are necessary for many semiconductor optoelectronic devices. The two most relevant factors for achieving carrier densities from doping in wurtzite group III nitrides are the electron (or hole) effective masses and the dielectric constant. In the hydrogenic model of the donor or acceptor, the binding energy is given by

$$E_b = 1Ry.\frac{m^*}{m_0}\frac{\varepsilon^2}{\varepsilon_0^2}$$

where 1 Ry = 13.6 eV is the Rydberg energy, m^* is the effective mass in the semiconductor, m_0 is the free electron mass, ε is the dielectric constant in the semiconductor, and ε_0 is the permittivity of free space. Effective masses of both electrons and holes tend to be higher in the group III nitrides compared to the arsenides and phosphides. Further, the dielectric constants are lower in the group III nitrides compared to the arsenides and phosphides. This means the binding energies for both donors and acceptors are higher. In fact, the acceptor binding energies can be 150 meV or even higher, meaning that they are deep acceptors. Since this is about six times the thermal energy at 300 K, the ionization ratios of acceptors can be as low as 0.01. That means getting good hole concentration and good *p*-type conductivity is a technological challenge. Special material and device tricks may be necessary to obtain the desired levels of *p*-type conductivity in the group III-Nitride semiconductors. Some papers discuss the acceptor ionization problem using k.p calculations. The k.p method is discussed in Chapter 4 in the context of low-dimensional structures, but we do not discuss the use of this method for evaluating acceptor ionization energies. The reason is that the hydrogenic model mentioned above gives most of the qualitative and some quantitative answers needed for our devices. It must be kept in mind that the k.p calculation for acceptor binding energy can be useful for specific devices.

References

1 de Carvalho, L.C. et al. (2011). Influence of exchange and correlation on structural and electronic properties of AlN, GaN, and InN polytypes. *Phys. Rev. B* (84) 195105.

2 Bungaro, C. et al. (2000). Ab initio Phonon Dispersions of wurtzite AlN, GaN and InN. *Phys. Rev. B* 61 (10).

3 Bulutay, C., Ridely, B.K. and Zakhleniuk, N.A. (2000). Full-band polar optical phonon scattering analysis and negative differential conductivity in wurtzite GaN. *Phys. Rev. B* 62 (23).

4 Jhalani, V.A. et al. (2020). Piezoelectric electron-phonon interaction from Ab initio dynamical quadrupoles: impact on charge transport in wurtzite GaN. *Phys. Rev. B 125*.

5 Ashcroft, N. and Mermin, N.D. (1976). *Solid State Physics*. Holt Rinehart and Winston.

6 Chuang, S.L. and Chuang, S. (2009). *Physics of Optoelectronic Devices*. Wiley.

7 Ridley, B. (2013). *Quantum Processes in Semiconductors*. Oxford University Press.

8 Ridley, B. (2011). *Electrons and Phonons in Semiconductor Multilayers*. Cambridge University Press.

3

Growth and Doping of Group III-Nitride Devices

Major Epitaxial Growth Methods

Growth Methods

The major growth methods for group III-Nitride semiconductor optoelectronic devices are molecular beam epitaxy (MBE), metal-organic chemical vapor deposition (MOCVD), and hydride vapor phase epitaxy (HVPE). These three methods have different attributes when it comes to control over layer thickness and uniformity, control over dislocation density and vacancies, uniformity of quantum well and wire widths, quantum dot (QD) dimensions, and growth rates.

Early designs of multi quantum-well light-emitting diodes (LEDs) using *c*-axis gallium nitride (GaN)/indium (In)GaN quantum wells faced the problem of serious red shift due to polarization-induced interface charges and built-in electric fields. Carrier screening cannot fully alleviate this problem since electrons and holes tend to be present in almost similar concentrations when the light-emitting diode is biased to give high optical output.

Molecular Beam Epitaxy (MBE)

MBE is a high vacuum growth method with layer-by-layer growth monitoring. The growth of each layer can be detected by reflection high energy electron diffraction (RHEED) patterns from the electrons emitted from a RHEED gun.

Components of MBE System

The major components needed to obtain epitaxial layers using MBE are:

1) Evaporation sources (effusion cells).
2) Plasma radio frequency (RF) sources (for the group V element, nitrogen for the devices considered in this book).
3) Shutters to control the molecular beam.
4) Susceptor with substrate.
5) Substrate heating apparatus with temperature control.
6) Ultra-high vacuum (UHV) chamber with pumps, with pressures reaching 10^{-9} torrs.

Large stainless-steel chambers with superfinished flanges are required to achieve the low pressures required for high-quality growth. To load the substrate, several chambers are connected that

Group III-Nitride Semiconductor Optoelectronics, First Edition. C. Jayant Praharaj.

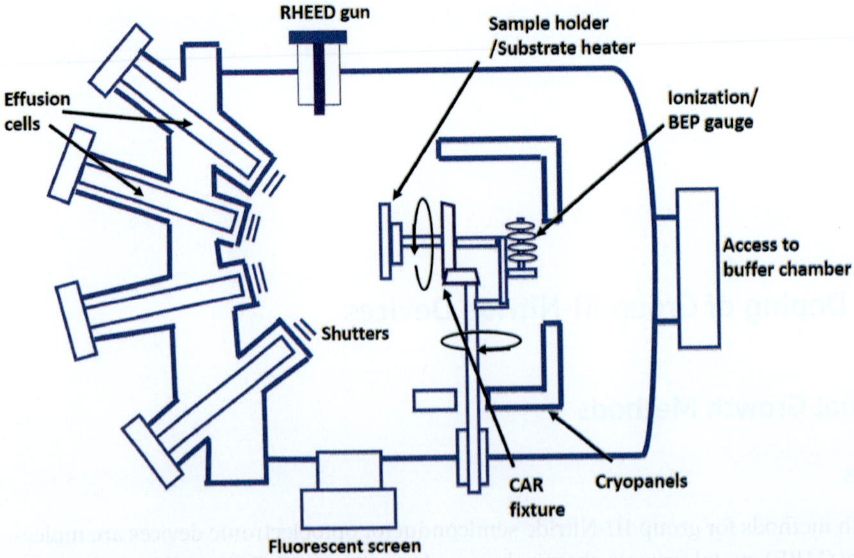

Figure 3.1 Schematic of an MBE system.

serve as gate chambers for the wafer. This helps maintain some amount of UHV in the chamber while exchanging substrates. Growth rates of one monolayer in about 10 seconds can be attained, enabling shutter control for monolayer growth or growth with monolayer precision. Figure 3.1 shows the basic configuration of an MBE growth system.

Effusion Cells

To achieve stable evaporation rates from the cell that depends only on cell temperature a Knudsen cell is used. This kind of cell has a small orifice that leads to a pressure in the cell equal to the equilibrium vapor pressure in the material. The rate at which atoms leave the Knudsen cell is given by the equation

$$\Gamma = \frac{p.A.N}{\sqrt{2\pi \mathrm{MRT}_c}} \tag{3.1}$$

and the deposition rate is given by

$$F = \frac{\Gamma}{\pi \mathrm{L}^2} \tag{3.2}$$

where L is the distance between the cell and the substrate. Typical cell temperatures during growth are between 250 °C and 1400 °C. Temperature control should be within 1°C accuracy to obtain stable ternary compositions. Cell materials can be molybdenum, tantalum, pyrolytic boron nitride (BN), or graphite to withstand the high cell temperatures.

The shutter controls the flow of atoms from the cell to the substrate. Correct shutter design is necessary to prevent unnecessary thermal coupling between the cell and the growth chamber.

Growth of III-nitride Nanowire Using MBE

The spontaneous growth of group III-Nitride nanowires typically happens in nitrogen-rich conditions on a silicon (Si) substrate. Anisotropy of surface properties like chemical potential and sticking coefficients play a crucial role in the growth of nanowires. Therefore, it is a diffusion-driven process.

Several studies have confirmed the existence of a thin amorphous interface constituted by a SiN*x* layer between the nanowire and Si substrate, which could suggest lattice constants different from the underlying substrate. This has been confirmed by the direct growth of GaN, InGaN/GaN dot-in-a-wire, and InN nanowire structures on the SiO*x* template. Importantly, it is noted that the nanowire structures exhibit comparable and/or better structural and optical properties compared to the nanowire structures grown directly on Si (Zhao et al. 2013a). These studies essentially lead to more choices of substrates for MBE-grown group III-Nitride nanowire structures and thus, render MBE-grown group III-Nitride nanowires promising candidates for a wide range of functional electronic and photonic devices. Recently, group III-Nitride nanowire LEDs grown directly on SiO*x* template has been transferred onto copper (Cu) and other flexible substrates to achieve enhanced performance (Nguyen et al. 2014).

InN and In-rich InGaN Nanowires Using MBE

The MBE growth of InN nanowire structures has been performed on Si and sapphire as well as reported extensively in the literature. A titanium (Ti) mask can be used on a GaN template to obtain InGaN nanowire growth. Figures 3.2 to 3.5 show characteristics of group III-Nitride nanowire structures.

AlGaN Nanowires Using MBE

AlGaN and aluminum nitride (AlN) nanowires can be grown on GaN templates similar to InGaN nanowires and with Si *n*-doping and magnesium (Mg) *p*-doping. Deep ultraviolet (UV) LEDs can be realized in these nanowire structures. Tuneable emission wavelengths covering UV A, B, and C bands can be achieved by MBE growth in N-rich conditions.

Stranski–Krastanow Growth Using MBE

The use of a wetting layer can lead to the formation of InGaN QDs, which leads to the Stranski–Krastanow (SK) growth regime. A typical wetting layer can be a monolayer of InN. QDs are useful

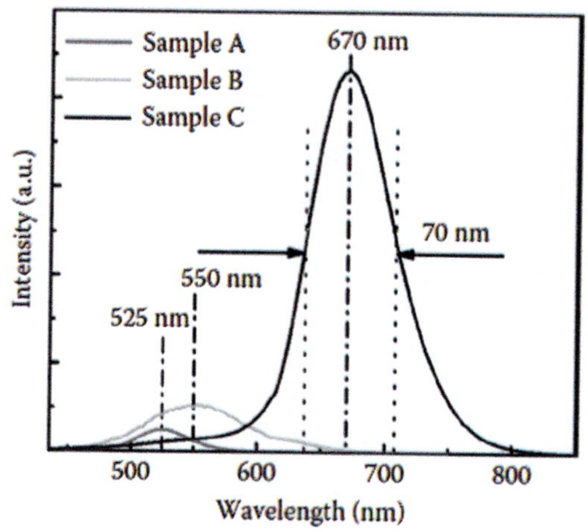

Figure 3.2 Room temperature (RT) photoluminescence emission of indium gallium nitride (InGaN) nanowires with three different indium mole fractions. S.Fan et al., APL Materials, 4, 076106, 2016a.

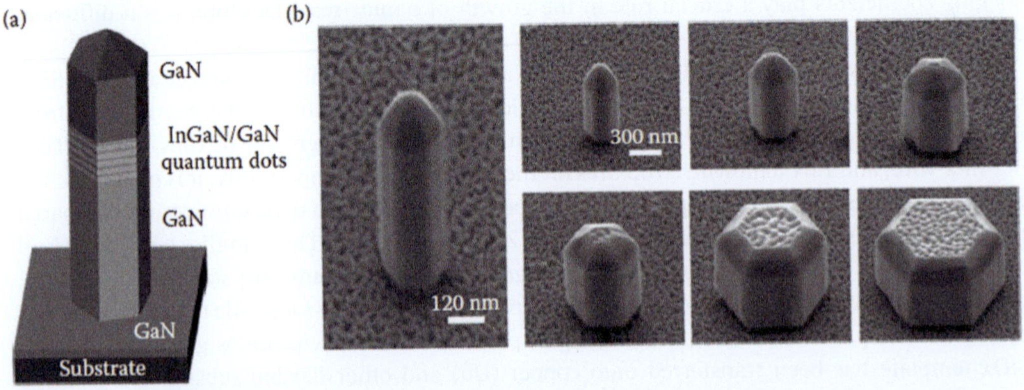

Figure 3.3 (a) InGaN/GaN quantum dot (QD) in nanowire grown on GaN template on sapphire substrate. (b) A scanning electron microscope (SEM) image of InGaN/GaN nanowires with various diameters. Y.H. Ra et al., Nano Letters, 2016.

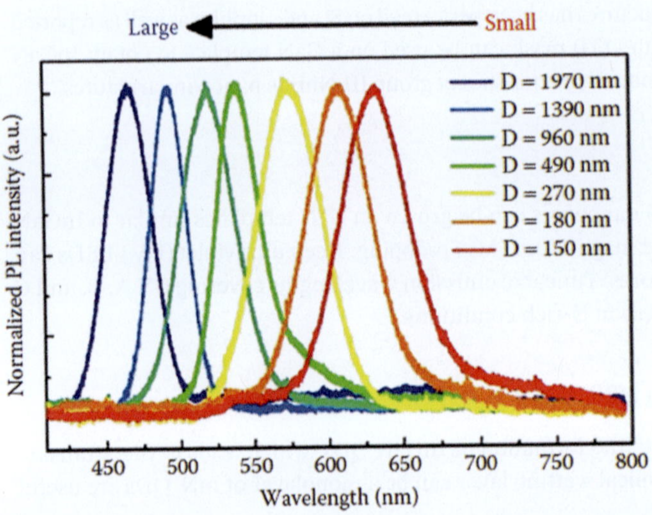

Figure 3.4 Normalized photoluminescence (PL) intensity of different diameter nanowires at RT. Y.H. Ra et al, Nano Letters, 2016.

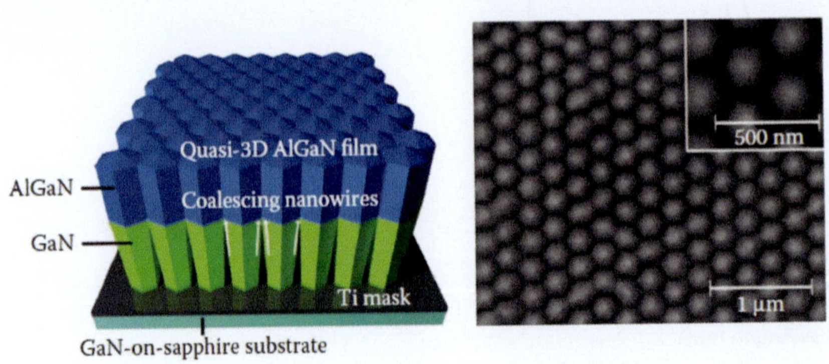

Figure 3.5 Coalescing AlGaN nanowires for LED fabrication. From B. Le et al, Advanced Materials, 2016.

for a variety of applications like LEDs and lasers. QDs tend to have lower built-in fields due to spontaneous and piezoelectric polarization. This can help reduce unwanted red shifts of emission, as discussed later in Chapter 4.

Quantum Dot (QD) Layer Growth and Dislocation Reduction for High Indium-content InGaN

Growing quantum wells with high In content often leads to bad material quality because of the large lattice mismatch between InGaN and GaN when the In content is large. Compared to quantum-well structures, QDs can help improve material quality of high In-content InGaN layer because of effective strain relaxation during the growth of QD structures rather than dislocations.

In Reference [2], temperature-dependent photoluminescence measurements were done to demonstrate high optical quality of QD layers grown by plasma-assisted MBE. The typical height and diameters of the dots were 2.3 nm and 28 nm. A low density of non-radiative recombination centers was found by temperature-dependent studies. The formation of QDs was proven by monitoring two-dimensional (2D)–three-dimensional(3D) morphological transformation through RHEED from streak to spot. Figures 3.6 and 3.7 show typical morphologies and luminescence of InGaN QDs.

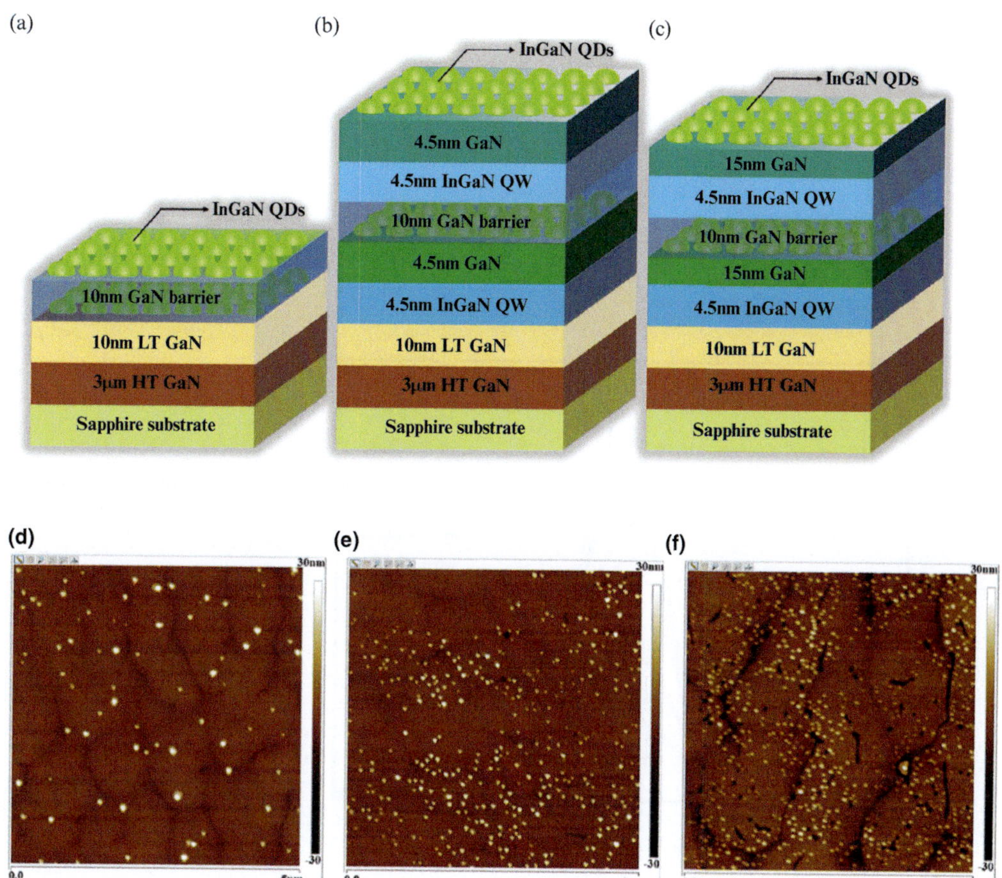

Figure 3.6 Stranski–Krastanow growth of InGaN QDs after Reference [1].

Strain-relaxed Templates for Achieving High-quality InGaN Layers for Red Wavelength Emission

Emission in the red wavelength requires high In-content InGaN, which typically has low quality due to lattice mismatch with GaN layers. Strain-relaxed templates (SRTs) with an InGaN decomposition layer have been used to fabricate red micro-LEDs in Reference [3]. References [4] and [5] also demonstrate various SRT-based growth of red LEDs using InGaN. In References [4] and [5], the SRTs consist of a thin layer with a high indium composition of an InGaN decomposition layer capped with GaN. The decomposition layer is then thermally decomposed to form voids during the high-temperature growth of either GaN or InGaN. After void formation, the subsequently grown InGaN layers show high levels of relaxation. This typically leads to high-quality InGaN layers that emit in the red part of the spectrum.

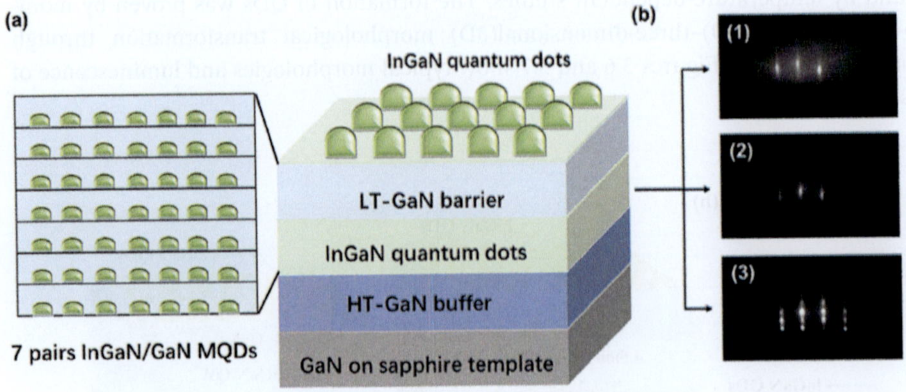

Figure 3.7(a) Schematic of InGaN QD growth. LT = low temperature; HT = high temperature.

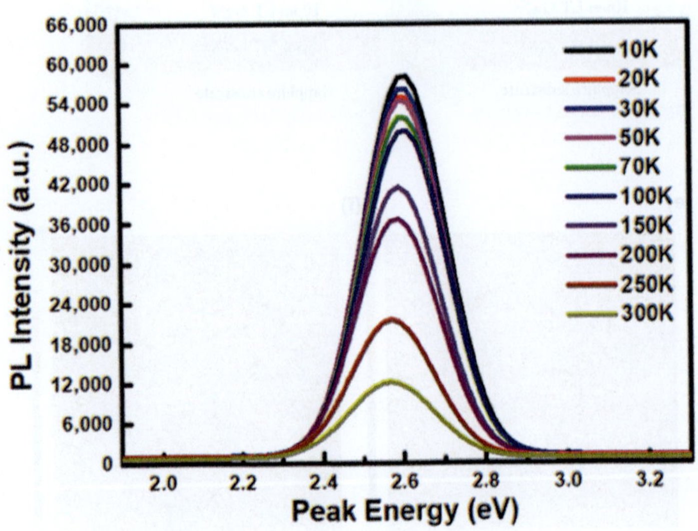

Figure 3.7(b) Typical photoluminescence from InGaN QDs, after Reference [1], used with permission.

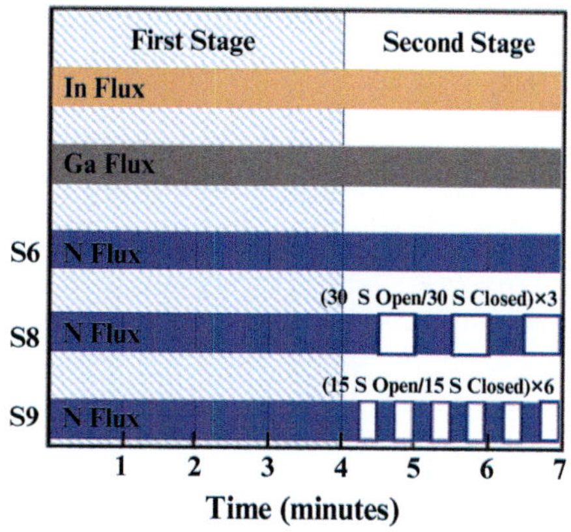

Figure 3.8 Nitrogen-modulation epitaxy stages. After Reference [6], used with permission.

Nitrogen-modulation Epitaxy Method Based on Stranski–Krastanov (SK) Mode

Nitrogen-modulation epitaxy has been explored in Reference [6] for the growth of high-quality QDs that emit in the red wavelength. In this method, there is a first stage where a wetting layer is grown in conditions of slight nitrogen enrichment. In the second stage, the nitrogen source is periodically turned on and off, while the metal sources of In and Ga are always maintained. Turning on the nitrogen source ensured N-rich conditions for growing QDs in SK mode, while when the nitrogen source was turned off, the reduction in the number of nitrogen atoms promotes the formation of 3D islands. The metal atoms left on the surface can diffuse laterally. This promotes In incorporation and a red shift of the emission wavelength. Figure 3.8 shows the typical stages of group III-Nitride nitrogen-modulation epitaxy.

MBE Growth of InGaN on Semi-polar and Non-polar Planes

Growth of InGaN nanorods (NRs) on patterned sapphire substrates (PSSs) has been demonstrated in Reference [7] for growth on semi-polar planes. This growth methodology uses sapphire patterning with wet etching to expose semi-polar planes of sapphire and uses these to obtain semi-polar plane nanorod growth. This overcomes problems with growing on large semi-polar planes related to poor heterointerface and dislocation densities. The connected, uniform, and obliquely aligned NRs were formed on the PSS with cylindrical arrays. We found that both pattern topography and crystallographic plane of the PSS significantly affect the morphology, dimension, and crystallographic orientation of InGaN NRs. It was clearly revealed that the PSS, which exhibits large-area crystallographic planes of a high Miller index with a well-organized step-terrace structure, facilitates the growth of ordered and uniform semi-polar NRs with high density.

Emission of photons from QDs contained in non-polar, semi-polar, and polar sections of pencil-like InGaN/GaN nanowires was demonstrated in Reference [7].

The QD types exhibit single photon emission with narrow emission linewidths and high degrees of linear optical polarization. This indicates high quality of the QD layers. Scanning transmission electron microscopy-cathodoluminescence (STEM-CL) was used to characterize the spatial distribution of photon emission.

Solar Water Splitting Using Group III-Nitride Nanowire Arrays (Photocatalysis Using Group III-Nitride Nanowires)

The indium gallium nitride (InGaN) alloy system covers the entire spectral range from 0.7 eV to 3.4 eV, which encompasses almost the entire solar radiation spectrum. InGaN has locations of

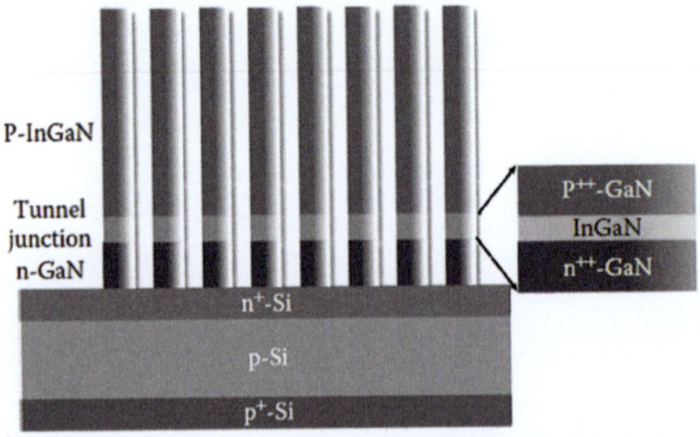

Figure 3.9 Water splitting vertical InGaN nanowire structure.

conduction-band minimum and valence-band maximum that are suitable for water splitting, as shown in Figure 3.9.

Vertical LED Structures

Vertical LED structures can be achieved by growing epilayers on Si substrates, Si substrate removal, roughening of GaN, and contact formation. The reader can find more details about vertical LED epitaxy and fabrication in recent literature on the subject.

Metal-organic Chemical Vapor Deposition

Precursors and Growth

The metallic elements need to be introduced in the form of metal-organic (MO) precursors. These are solid or liquid and have high vapor pressures. Tri-methyl-gallium ($TMGa{=}Ga[CH_3]_3$) is a liquid at RT with a boiling point of 55.8 °C and a vapor pressure of 8560 Pa at 0 °C. The precursor is stored in a bubbler at a stabilized temperature with a specific vapor pressure. A carrier gas (typically hydrogen) is flown through the bubbler. The carrier gas gets saturated according to the precursor vapor pressure. The mass flow rate of the precursor is determined by the equation

$$n_{MO} = \frac{r_{H_2} p_{vap}(T) p_o}{p_{Bubbler} R T_o} \tag{3.3}$$

where n_{MO} is the transport rate in moles per minute, and r_{H_2} is the flow of carrier gas through the bubbler in $m^3\ min^{-1}$. p_o is the standard pressure of 10^5 Pa, $p_{bubbler}$ is the pressure in the bubbler, R is the gas constant, and T_o is the standard temperature of 15°C. The MO precursors should be chemically stable in gas phase when mixed with other precursor gases, should undergo thermal cracking at hot substrates to release the target element, should have high purity so as not to contaminate the semiconductor, should release only the wanted element onto the substrate, and should have good physical and chemical safety standards (non-pyrophoric, nonexplosive, nontoxic). Figure 3.10 shows a typical MOCVD reactor configuration.

The smallest organic molecules containing the metal element are preferable from vapor pressure and purity considerations, for example, TMGa, TMAl, and TMIn. Sometimes ethyls like tri-ethyl-gallium $Ga(C_2H_5)_3$ can be used. For aluminum deposition, Al and TMAl are very reactive and lead

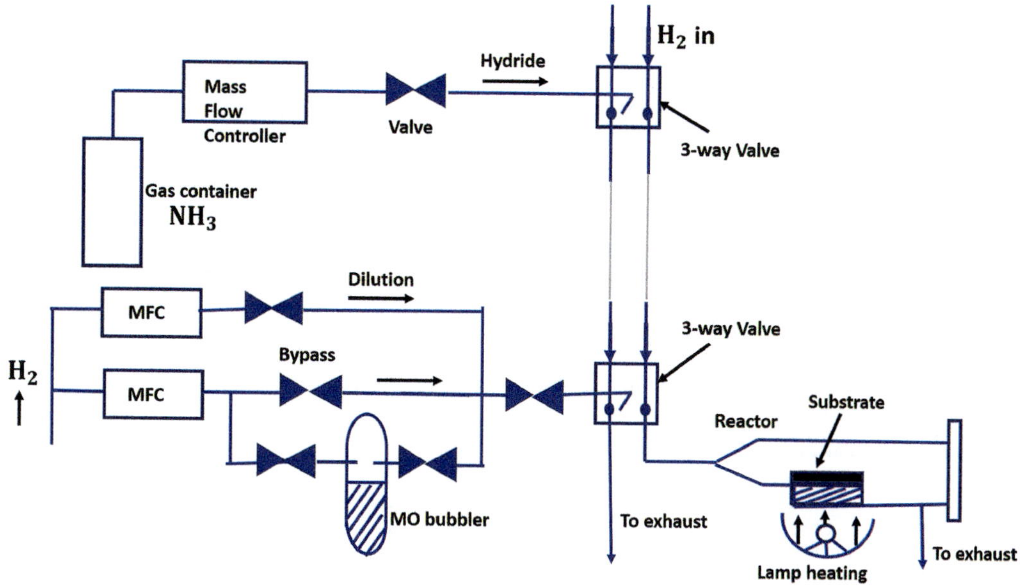

Figure 3.10 A typical MOCVD reactor configuration.

to potential oxygen contamination. Furthermore, the high bond strength of the Al–C bond can potentially lead to C uptake. TMIn has melting point of 88°C and has to be used in solid form as a powder. It sometimes necessitates special bubble construction.

The reaction leading to semiconductor layer deposition for TMGa is given by

$$Ga(CH_3)3 + NH_3 \rightarrow GaN + 3CH_4 \tag{3.4}$$

As a function of the susceptor temperature, different regimes can be identified. At low temperature, the growth rate is limited by the chemical reaction rate and is given by an Arrhenius rate equation

$$k = Ce^{-\frac{E_A}{kT}} \tag{3.5}$$

At medium temperatures, the growth rate becomes limited by group III precursor supply and is not sensitive to temperature. At very high temperatures, desorption and dissociation of the crystal can limit growth rate. The group V precursor gases are supplied in excess because hydrides are stable, and group V elements desorb easily from the hot surface without excess partial pressure. The reader can find further details in Reference [8]. Figures 3.11 and 3.12 show typical growth-limiting steps and boundary layer physics.

Adduct formation during growth provides pathways to the formation of final products in MOCVD. In ammonia (NH_3) rich conditions, an adduct where one hydrogen of NH_3 bonds with C from $(CH_3)_3Ga$ leads to a reaction pathway with subsequent elimination of CH_4.

Incorporation of In in InGaN growth is favored by higher growth pressure. These conditions, however, result in In droplets that must be evacuated before growing the GaN or AlGaN barrier

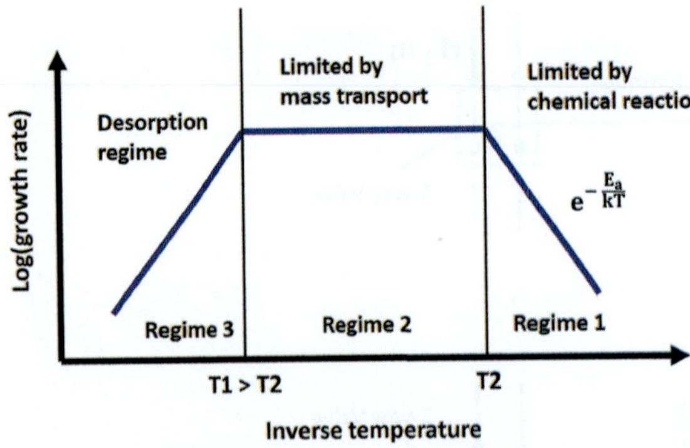

Figure 3.11 MOCVD growth-rate limiting steps at different temperatures.

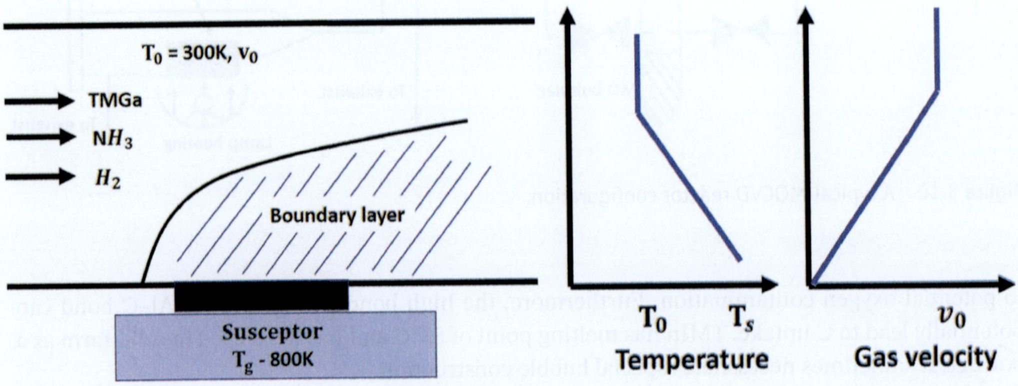

Figure 3.12 Boundary layer details for MOCVD growth.

layer. This is done by a temperature ramp to remove the droplets before ramping down again to start growing the barrier layer.

Carbon incorporation in epilayers during MOCVD growth can affect device properties. This is because C forms a deep level in nitride semiconductors. It can compensate donors and acceptors and lower layer mobilities significantly. It can also serve as a highly localized scattering center and contribute to reduced mobility. A high group V/III ratio, higher growth pressure, or higher temperature tends to reduce carbon incorporation (Reference [9]). At a constant group V/III ratio, carbon incorporation is independent of the growth rate.

The conditions for carbon incorporation affect the growth conditions that can be chosen for Mg doping. The active regions of visible optical devices are made of thermally unstable InGaN. Therefore, the thermal budget associated with growing a Mg-doped *p*-clad layer needs to be reduced. Furthermore, Mg diffusion from the barrier layer to the optically emitting layer is reduced at lower temperatures. However, a lower temperature leads to higher C incorporation, and that makes it necessary to do a careful optimization to get functioning devices.

Doping of Group III Nitrides during MOCVD Growth

Dopants can be supplied as MO liquids or as hydrides. Si can replace the group III element and act as a donor. It can be introduced using silane (SiH_4). Mg, calcium (Ca), or strontium (Sr) can replace the group III element and act as an acceptor. Mg can be introduced using bis-cyclo-pentadienyl-Mg (Cp_2-Mg). Iron (Fe) can be used to get deep levels for semi-insulating behavior. $Fe(C_5H_5)_2$ can be used as the precursor.

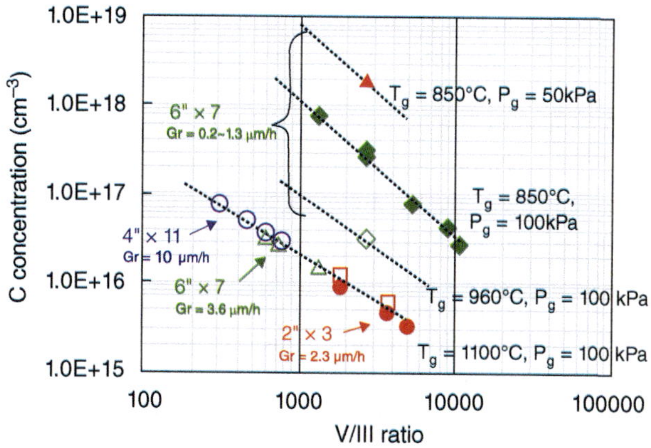

Figure 3.13 Carbon concentration during MOCVD growth of GaN at different growth conditions. From Reference [9], used with permission.

In Situ Characterization

In situ characterization techniques include reflection coefficient measurement, reflection anisotropy spectroscopy, pyrometric true temperature evaluation, and laser-based curvature measurement.

Gas-phase Chemistry Inhibition for Group III-Nitride MOCVD Growth

Nakamura et al. were the first to report enhanced growth quality using reduced parasitic gas-chemistry and better growth rates. They had to do a special reactor design to suppress gas-phase pre-reaction in the boundary layer.

Nakamura pioneered a design where the MOs and NH_3 precursors were delivered from a side quartz tube and pushed vertically against the surface using high flow rates of N_2 and H_2 carrier gas. With this method, they were able to get atmospheric pressure MOCVD with electron mobility of $900\,cm^2V^{-1}sec^{-1}$ at RT and $3000\,cm^2V^{-1}sec^{-1}$ at 77 K. Methods that keep the MO and NH_3 separate until they reach the heated surface help to minimize the parasitic gas-phase reactions. Rotating the wafer carrier at high speed produces a viscous drag on the vertical gas flow and pulls it horizontally across the wafer surface and thins the boundary layer, thereby achieving lower rates of parasitic gas phase. In horizontal reactors, one way to avoid gas-phase parasitic reactions is to reduce the reagent residence time by increasing the flow rates through the reactor. In close-coupled showerhead and horizontal reactors, the degree of gas-phase particulate formation could be reduced by heating the top reactor wall.

GaN Heteroepitaxy Using MOCVD

Growing GaN directly on sapphire tends to produce a very rough morphology. The use of AlN nanolayers on top of the sapphire substrate produced very uniform layers. Sapphire is chemically stable at the high temperatures required for GaN deposition and is not corroded by NH_3 or H_2. It can be obtained in chemically pure forms, is polishable to a smooth finish, and is cheap.

GaN and AlN Nanolayers for Nitride MOCVD

Low-temperature AlN nanolayers on sapphire, low-temperature GaN nanolayers on sapphire, and AlN nanolayers on silicon carbide (SiC) help achieve group III-Nitride layers with better morphology.

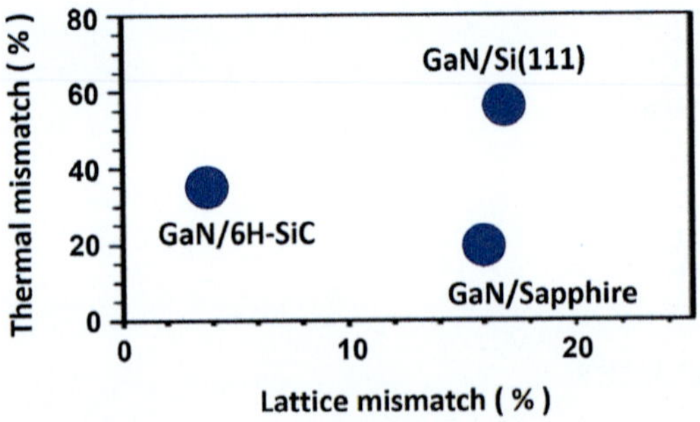

Figure 3.14 Lattice mismatch and thermal mismatch between GaN and some substrates.

In Situ Monitoring

The MOCVD growth temperature is controlled using proximity thermocouples, optical pyrometry, or applied heater power. Optical reflectance can be used as an in situ diagnostic. Stress monitoring can be done by measuring wafer curvature using single or multiple laser beams. InGaN LEDs on Si (111) could be grown by continuous strain monitoring and by inserting AlN and AlGaN strain-relieving layers. Figure 3.14 shows thermal mismatch and lattice mismatch for GaN growth on various substrates.

MOCVD Growth of InGaN on Semi-polar and Non-polar Planes

Comparison of MOCVD growth of InGaN on polar and semi-polar planes in Reference [10] shows that In incorporation is easiest for the polar plane and more difficult for the semi-polar planes. As will be discussed in later chapters, polar plane growth has more polarization-induced built-in electric fields, which leads to the degradation of emission characteristics.

In Reference [11], white emission was achieved using blue and yellow quantum wells in combination. Polarized monolithic white semi-polar LEDs with a top blue quantum well (QW) and a bottom yellow QW were grown on large-area (20–21) semi-polar GaN/sapphire template, as shown in Figure 3.15.

Quantum dashes or QDs of InN were grown in Reference [12] on 20–2–1 GaN substrates. This demonstration is important because it shows the possibility of achieving InGaN QD growth on semi-polar and non-polar planes. While InGaN QDs on polar *c*-planes lead to improved emission and device efficiencies in many studies, InGaN QDs on non-polar and semi-polar planes are of fundamental interest because they can potentially lead to new device physics because there is small or no built-in polarization field along the growth direction.

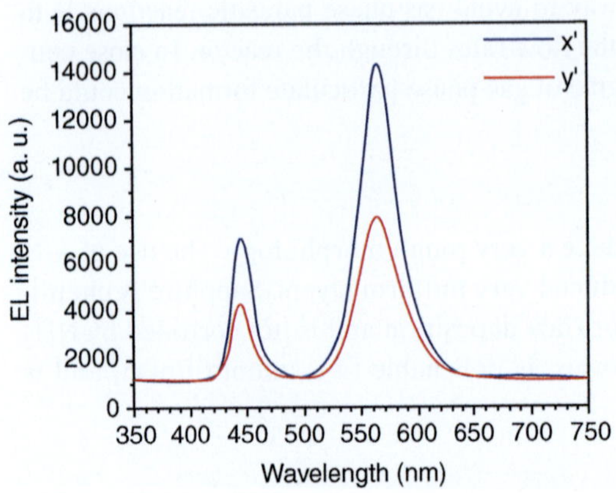

Figure 3.15 Electroluminescence of combined blue and yellow quantum wells using MOCVD growth on a semi-polar GaN/sapphire template. After Reference [11], used with permission.

Long-wavelength and blue InGaN LEDs have been demonstrated in Reference [13] on semi-polar planes using MOCVD growth on patterned GaN/sapphire templates. Epitaxy on patterned templates typically leads to lower stacking fault density and lower dislocation density compared to growth on non-patterned substrates.

In Situ and Implant Doping

In situ doping takes place during MBE or MOCVD growth by adding dopant precursors during growth. Typically, in situ doping causes fewer defects compared to implant doping since implant doping depends on high energy dopant ions. Implant doping should be followed by annealing to activate the dopants and to decrease the number of defects.

Dislocations and Point Defects

Dislocations are typically present when the layer grown is thicker compared to critical thickness. For layers with thickness less than the critical thickness, the epitaxial layer adjusts its lattice constant to that of the underlying substrate. However, as the thickness increases, the strain energy present in the epitaxial layer can exceed the energy associated with formation of dislocations. This critical thickness has been calculated in the literature. For example, there is a Matthews–Blakeslee critical thickness limit that determines when dislocation multiplication takes place for many materials. Dislocations are typically considered bad for optoelectronic devices because they provide deep states in the bandgap for non-radiative recombination of carriers. This non-radiative recombination adversely affects the efficiency of light emitters as well as photovoltaic cells and photodetectors.

Dopant-induced Defects

Implanted layers are typically used in electronic devices, while optoelectronic devices tend to use in situ doping. Implanted doping can lead to dopant-induced vacancies. These vacancies, like dislocations, provide deep states in the bandgap that lead to unwanted non-radiative recombination.

Substrates and Growth

Growth of Group III Nitrides by MOCVD on Sapphire Substrates

Sapphire (Al_2O_3) has a hexagonal crystal structure, which makes it a top choice for growth of wurtzite group III-Nitride semiconductors. Furthermore, sapphire is a high chemical and thermal stability material. It has a melting temperature of 2323 K (2030°C) and maintains its stability at very high temperatures of GaN buffer epitaxial growth, which may need to be higher than 1000°C. Sapphire is very resistant to chemical attack by a wide range of chemicals even at elevated temperatures. H_2 is used as a carrier gas in MOCVD, and it can cause thermal etching of the substrate at high temperatures. The high chemical stability of sapphire makes it ideal for the growth of many different device layers.

Growth of Group III Nitrides by MOCVD on SiC Substrates

SiC has several polytypes of the wurtzite crystal structure form. Of these, 4H- and 6H-SiC substrates, mostly on a Si-face surface, are used for the heteroepitaxy of wurtzite GaN. The in-plane lattice constants (a = 3.073 Å and a = 3.081 Å for 4H- and 6H-SiC) are much closer to GaN and other group III nitrides than that of sapphire. The in-plane lattice mismatch between GaN and SiC is significantly smaller (3.5%) than that between GaN and sapphire (14%). This lattice mismatch is further improved using an AlN buffer layer between the GaN layer and the SiC substrate since the lattice constant of AlN is between the values of GaN and SiC. The lattice mismatch between AlN and SiC is 2.45% for AlN on 6H-SiC and GaN on AlN. For the case of GaN directly grown on SiC substrates, achieving layer-by-layer 2D growth is difficult, possibly due to a wetting problem of GaN on SiC or the onset of SK growth mode. Like in the case of sapphire, the thermal expansion coefficient mismatch between SiC and GaN must be kept in mind while designing growth strategies. The reader can find details of group III-Nitride epitaxy for devices in References [14–30].

Non-polar and Semi-polar Growth of Group III Nitrides on Sapphire and SiC Substrates

The non-polar (1120) or *a*-plane GaN is typically grown on (1102) *r*-plane sapphire substrates and on *a*-plane sapphire and SiC substrates. The non-polar (1010) *m*-plane GaN is grown on *m*-plane sapphire and SiC substrates. The reader can find details of non-polar and semi-polar growth in References [31–37].

Gallium Nitride Growth on Silicon Substrates

Growth on silicon substrate is of great interest because of the lower cost of wafers. Typically, wurtzite GaN and other wurtzite nitrides are grown on (111) silicon. The *a*-axis lattice parameter of GaN is 3.18 Å while the effective lattice constant of (111) Si is 3.84 Å. The lattice mismatch is 16.9%, which is large and gives a high density of dislocations. The strain of GaN on Si is tensile, while that on sapphire is compressive, which affects buffer layer requirements for reducing dislocations.

Typically, GaN (0001) grows on Si (111) and GaN (1120) grows on Si (110). However, several studies have tried to grow non-polar axis GaN on Si (111) with lower lattice constant mismatch.

For MOCVD growth on Si, the hydrogen in the ambience can cause some chemical reactions to occur at the substrate surface. These can affect roughness of the grown GaN layer. Also, Si diffuses quite fast and it can dope the GaN or AlN layer and affect overall device performance when further epilayers are grown on the GaN or AlN.

Thermal mismatch is a serious concern when GaN is grown on Si substrate. The thermal expansion coefficient of gallium nitride is $5.59 \times 10^{-6}\,K^{-1}$ for gallium nitride in the *a*-axis direction and $3.59 \times 10^{-6}\,K^{-1}$ for silicon. This can lead to serious problems like cracking when the wafer cools down to RT after growth.

Buffer layers can be used to address the problem of large lattice mismatch between Si and GaN and the tensile strain. AlN buffer layers, which convert the strain in the GaN from tensile to compressive, have been explored to improve GaN layer quality. Other buffer layers are GaAs, AlAs, boron phosphide (BP), and zinc oxide (ZnO). One advantage of AlN as a buffer layer is the fact that it has the same type of crystal structure as GaN, which helps address the dislocation problem. The Si is trapped as silicon nitride (SiN) in the AlN layer and the GaN layer remains unaffected. Growth temperature affects the quality of the final GaN layer when AlN is used as the buffer layer. Typically, low-temperature (LT)-AlN prevents cracking of the epitaxial GaN layer and high temperature (HT)-AlN buffer layer results in better crystal quality of the GaN layer.

Superlattice layers have also been explored to reduce GaN crystal quality when grown on Si. One way is the growth of GaN/AlN superlattices on the AlN buffer on Si substrate before growing the GaN device layer. Raman scattering can be used to obtain the strain level in the GaN layer and X-ray crystallography is used to monitor the crystal quality. Typically, 10–20 superlattice layers are used before growing the GaN layer. Raman scattering shows lower stress and X-ray diffraction shows better crystal quality with the use of the superlattice interlayers. Dislocation density reduction is demonstrated, and LED device performance has also been seen. Parameters like power efficiency and quantum efficiency improve when the superlattice interlayers are used. The use of the superlattice interlayers involves higher growth complexity, but the improvement in device quality justifies it in many cases. The use of a nucleation layer before growth of the buffer layer is another way to improve GaN layer quality in many cases. Figure 3.16 shows typical LED characteristics for nitride growth on Si substrate.

Indium gallium nitride (InGaN) epilayers have also been used to improve device quality, especially in LEDs. Figure 3.13 shows the effect of InGaN superlattice prelayer on LED device performance for devices built on silicon substrates. Internal quantum efficiency (defined in Chapter 5) improves with InGaN superlattice prelayer and the efficiency droop is reduced.

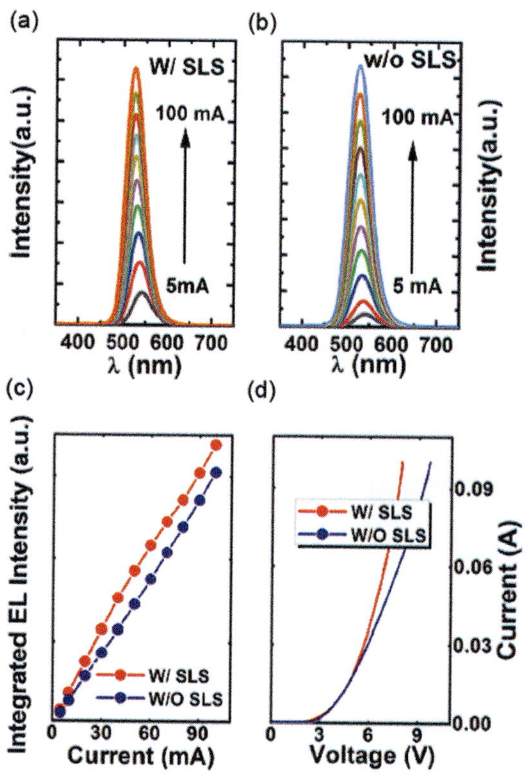

Figure 3.16 LED characteristics for green LED on Si substrate with and without InGaN superlattice prelayer (from Reference [38], used with permission).

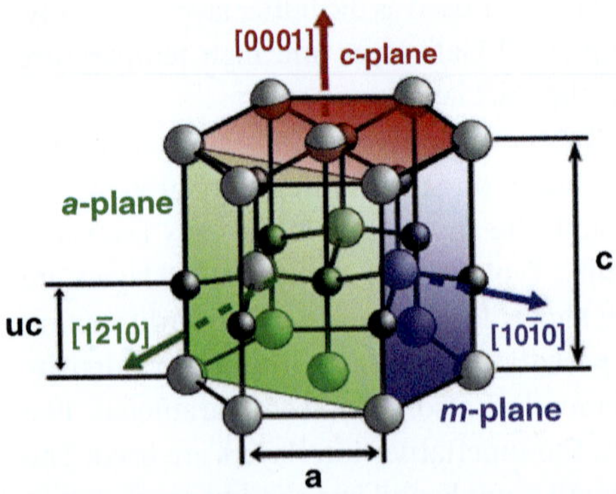

Figure 3.17 Wurtzite crystal structure showing non-polar *a*- and *m*-planes.

Non-polar and Semi-polar Growth of Group III Nitrides on Si Substrates

GaN (1101) growth on Si (001) and GaN (1122) growth on Si (113) are some of the typical combinations used to obtain non-polar or semi-polar growth on Si substrates. Figure 3.17 shows various polar and non-polar planes. Specialized etching and masking techniques can become necessary to obtain the right substrate face orientation to obtain non-polar face growth. Since non-polar and semi-polar growth are often needed to obtain LEDs of high efficiency and lasers with low threshold current and high efficiency, the optimization of non-polar growth on Si attracts a lot of research effort. One of the major concerns with growth on non-polar and semi-polar nitrides on silicon is the dislocation density. Since most of the early efforts were focused on the polar *c*-plane growth, the growth optimization involving temperature during growth, flux rates, etc. were done for *c*-plane as well. Getting high-quality InGaN and other nitride layers with low dislocation density on semi-polar and non-polar plane still needs a lot more optimization to obtain high-quality device layers. Figure 3.18 shows non-polar LEDs on Si substrates.

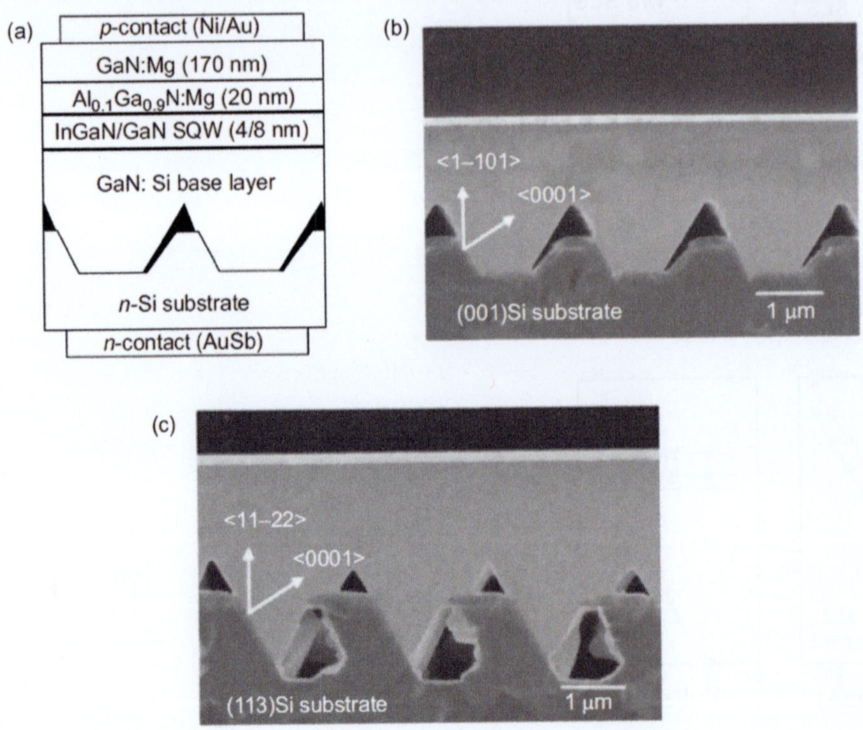

Figure 3.18 Typical non-polar LEDs on Si substrates (after Reference [39], used with permission).

Group III-Nitride Nanowire Growth on Si Substrates

Since nanowires often give high efficiency LEDs and lasers due to reduced piezoelectric fields, the combination of nanowire growth with cost-effective Si substrates is of great interest. Several studies have tried to optimize group III-Nitride nanowire growth on Si substrates with different temperature and pressure conditions and use of wetting layers. The nanowire lateral dimensions often tend to be on the order of 50–100 nm when nanowire growth is done on Si. However, device optimization often requires the lateral dimensions of the nanowire to be lower so that the piezoelectric (and spontaneous polarization) fields can be reduced. Therefore, more optimization is needed to get good device quality nanowires on Si substrates. Optimization of the wetting layer can provide a way to getting high device quality nanowires on Si substrates. Figure 3.19 shows characterization of typical group III-Nitride nanowires on Si substrates. The reader can find more details about gorup III nitride growth in reference [42].

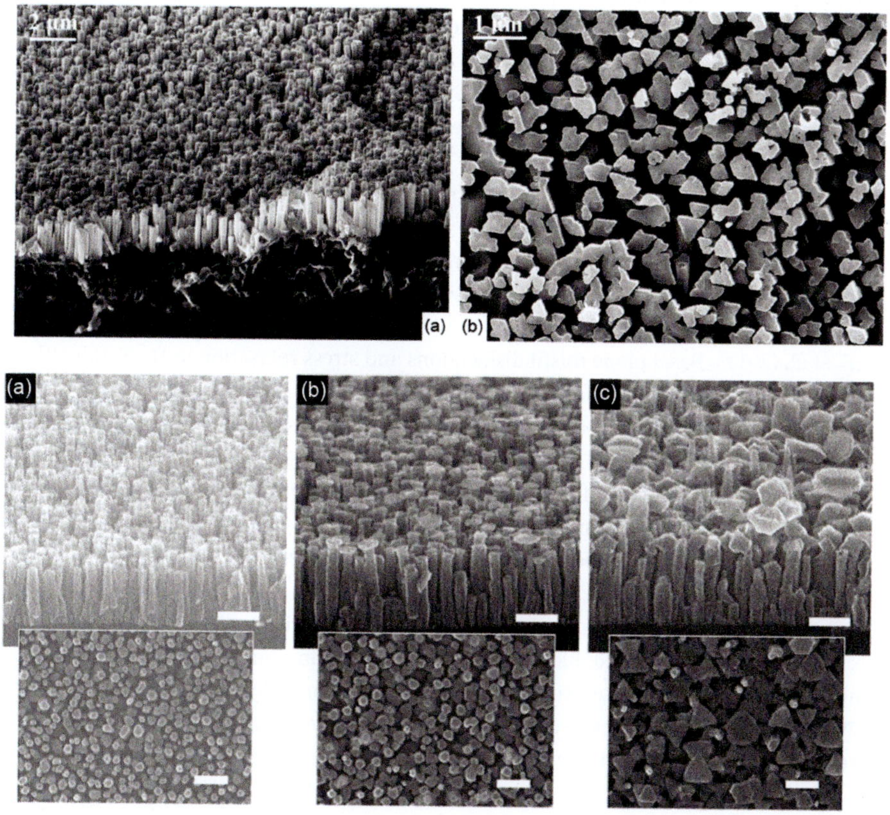

Figure 3.19 Typical nanowires obtained on Si substrates. From Reference [40] (top) and Reference [41] (bottom), used with permission.

References

1 Wang, J. et al. (2018). Abnormal Stranski–Krastanov mode growth of green InGaN quantum dots: morphology, optical properties, and applications in light-emitting devices. *Appl. Mater. Interfaces* 11 (1).

2 Zhang, X. et al. (2021). Investigations of micromorphology and carrier recombination dynamics for InGaN/GaN multi-quantum dots grown by molecular beam epitaxy. *Crystals* 11 (11).

3 Wong, M. et al. (2022). Low forward voltage III-Nitride red Micro-light-emitting diodes on a strain relaxed template with an InGaN decomposition layer. *Crystals* 12 (5).

4 Chan, P., DenBaars, S.P., and Nakamura, S. (2021). Growth of highly relaxed InGaN pseudo-substrates over full 2-in. wafers. *Appl. Phys. Lett.* 119 (13).

5 Chan, P., Rienzi, V., Lim, N. et al. (2021). Demonstration of relaxed InGaN-based red LEDs grown with high active region temperature. *Appl. Phys. Express* 14 (10).

6 Zhang, X. et al. (2022). Explorations on growth of Blue-Green-Yellow-Red InGaN quantum dots by Plasma-assisted molecular beam epitaxy. *Nanomaterials* 12 (5).

7 Shen, J. et al (2018). Crystallographic plane and topography-dependent growth of semipolar InGaN nanorods on patterned sapphire substrates by molecular beam epitaxy. *Nanoscale* 46.

8 Scholz, F. (2018). *Compound Semiconductors_Physics, Technology and Device Concepts*. Pan Stanford Publishing.

9 Philip, M.R. et al. (2016). High efficiency green/yellow and red InGaN/AlGaN nanowire light-emitting diodes grown by molecular beam epitaxy. *J. Crys. Growth* 2 (2).

10 Dinh, D. et al. (2016). MOVPE growth and indium incorporation of polar, semipolar (11 ‾ 22) and (20 ‾ 21) InGaN. *Phys. Status Solidi* 253 (1).

11 Khouri, M. et al. (2019). Polarized monolithic white semipolar (20–21) InGaN light-emitting diodes grown on high quality (20–21) GaN/sapphire templates and its application to visible light communication. *Nano Energy* 67.

12 Chan, P. et al. (2021). Growth by MOCVD and photoluminescence of semipolar (2021) InN quantum dashes. *J. Cryst. Growth* 563.

13 Li, H. et al. (2020). Development of efficient semipolar InGaN long wavelength light-emitting diodes and blue laser diodes grown on a high quality semipolar GaN/sapphire template. *J. Phys.: Photonics* 2 (3).

14 Romanov, A.E. et al. (2011). Basal plane misfit dislocations and stress relaxation in III-nitride semipolar heteroepitaxy. *J. Appl. Phys.* 109: 103522.

15 Dadgar, A., Schulze, F., Wienecke, M. et al. (2007). Epitaxy of GaN on silicon – impact of symmetry and surface reconstruction. *New J. Phys.* 9.

16 Wong, M. et al. (2018). High efficiency of III-nitride micro-light-emitting diodes by sidewall passivation using atomic layer deposition. *Opt. Express* 26 (16).

17 Ni, X., Wu, M., Lee, J. et al. (2009). Nonpolar m-plane GaN on patterned Si(112) substrates by metalorganic chemical vapor deposition. *Appl. Phys. Lett.* 95 (11).

18 Ra, Y.H. et al. (2016). Full color single nanowire pixels for projection displays. *Nano Lett.* 16 (7).

19 Chiu, C., Lin, D., Lin, C. et al. (2011). Reduction of efficiency droop in semipolar (1101) InGaN/GaN light emitting diodes grown on patterned silicon substrates. *Appl. Phys. Exp.* 4 (3).

20 Le, B. et al. (2016). Controlled coalescence of AlGaN nanowire arrays: anarchitecture for nearly dislocation-free planar ultraviolet photonic device applications. *Adv. Mat.* 28 (38).

21 Shen, S. et al. (2019). Semi-polar InGaN-based green light-emitting diodes grown on silicon. *Phys. Status Solidi A* 217: 1900654.

22 Tendille, F., De Mierry, P., Vennéguès, P. et al. (2014). Defect reduction method in (11–22) semipolar GaN grown on patterned sapphire substrate by MOCVD: toward heteroepitaxial semipolar GaN free of basal stacking faults. *J. Cryst. Growth* 404.

23 Vennéguès, P., Tendille, F., and De Mierry, P. (2015). Study of defect management in the growth of semipolar (11–22) GaN on patterned sapphire. *J. Phys. D: Appl. Phys.* 48 (32).

24 Khoury, M. et al. (2019). Demonstration of electrically injected semipolar laser diode grown on low cost and scalable sapphire substrates. *ACS Appl. Mater. Interfaces* 11 (50).

25 Haller, C., Carlin, J.F., Jacopin, G. et al. (2017). Burying non-radiative defects in InGaN underlayer to increase InGaN/GaN quantum well efficiency. *Appl. Phys. Lett.* 11 (26).

26 Gacevic, C. et al. (2017). Emission of linearly polarized single photons from quantum dots contained in nonpolar, semipolar and polar sections of pencil-like InGaN/GaN nanowires. *ACS Photonics* 4 (3).

27 Davies, M.J. et al. (2015). A study of the inclusion of prelayers in InGaN/GaN single- and multiple-quantum-well structures. *Phys. Status Solidi B* 252 (5).

28 Song, J. et al. (2019). Elimination of stacking faults in semipolar GaN and light-emitting diodes grown on sapphire. *ACS Appl. Mater. Interfaces* 11 (36).

29 Armstrong, A.M. et al. (2015). Defect-reduction mechanism for improving radiative efficiency in InGaN/GaN light-emitting diodes using InGaN underlayers. *J. Appl. Phys.* 117 (13).

30 Niu, N. et al. (2006). Improved quality of InGaN/GaN multiple quantum wells by a strain relief layer. *J. Cryst. Growth* 286 (2).

31 Song, J., Choi, J., Xiong, K. et al. (2017). Semipolar (20–2–1) GaN and InGaN light-emitting diodes grown on sapphire. *ACS Appl. Mater. Interfaces* 9 (16).

32 Fan, S. et al. (2016). An In0.5Ga0.5N nanowire photoanode for harvesting deep visible light photons. *APL Mat.* 4 (7): 076106.

33 Tanikawa, T., Hikosaka, T., Honda, Y. et al. (2008). Growth of semi-polar (11e22)GaN on a (113)Si substrate by selective MOVPE. *Phys. Stat. Sol. C* 5 (9).

34 Rashidi, A., Monavarian, M., Aragon, A. et al. (2018). Nonpolar m-Plane InGaN/GaN micro-scale light-emitting diode with 1.5 GHz modulation bandwidth. *IEEE Electron Device Lett.* 39 (4).

35 Li, J. and Zhang, Z.Q. (eds.) (2019). *Light Emitting Diiodes_Materials, Processes, Devices and Applications*. Springer International Publishing.

36 Törmä, P.T. et al. (2008). Effect of InGaN underneath layer on MOVPE-grown InGaN/GaN blue LEDs. *J. Cryst. Growth* 310.

37 Poyiatzis, M. et al. (2019). Monolithically integrated white light LEDs on (11–22) semi-polar GaN Templates. *Sci. Rep.* 9.

38 Zhao, X. et al. (2020). Influence of an InGaN superlattice pre-layer on the performance of semi-polar (11–22) green LEDs grown on silicon. *Sci. Rep.* 10.

39 Hikosaka, T., Tanikawa, T., Honda, Y. et al. (2008). Fabrication and properties of semi-polar (1101) and (1122) InGaN/GaN light emitting diodes on patterned Si substrates. *Phys. Stat. Sol. C* 5 (6).

40 Reznik, R.R. et al. (2020). MBE synthesis and properties of GaN NWs on SiC/Si substrate and InGaN nanostructures on Si substrate. *J. Phys.* 1537.

41 Gridchin, V.O. et al. (2021). Morphology transformation of InGaN nanowires grown on Si substrate by PA-MBE. *J. Phys.: Conf. Ser.* 2086.

42 Bi, W., Kuo, H., Ku, P. and Shen, B. eds. (2018). *Handbook of GaN Materials and Devices*. CRC Press.

4

Optical Properties of Low-dimensional Structures in Group III Nitrides

Quantum Wells, Quantum Wires, and Quantum Dots

While bulk semiconductors allow the formation of devices, the formation of low-dimensional structures allows properties like energy spectra to be modified according to need. Often, low-dimensional structures like quantum wells, quantum wires, quantum dots (QDs), and superlattices can lead to novel device phenomena and improved device characteristics. One of the most significant changes is that the density of states (DOS) for electrons and holes gets modified in quantum wells, wires, and dots. This leads to a modification of absorption coefficients, gain coefficients, etc. Typically, laser and light-emitting diode (LED) device characteristics like threshold currents can be improved using quantum wells and quantum wires. QDs also provide similar improvements, however, some further novel physical effects come into play that must be understood for optimum device design.

Common Quantum Wells in Group III Nitrides

The most common and technologically important quantum wells in the group III-Nitride material system are of the gallium nitride (GaN)/InGaN/GaN variety [1–4]. Since indium gallium nitride (InGaN) allows the bandgap to be varied over the wide range of 0.7 eV to 3.43 eV, the photon energies associated with quantum-well level transitions can be correspondingly varied over a wide range, as will be shown. However, the quantum wells of widest interest provide emission in the blue wavelength region of the optical spectrum.

Other kinds of quantum wells with different compositions have been studied as well. They are of potential interest for realizing devices, particularly those with optical emission in the higher energy or lower wavelength regime compared to InGaN quantum wells.

Multi-quantum Well and Staggered Quantum Wells

Multiple quantum wells often allow better optimization of device properties. Further, staggered quantum wells, where the InGaN composition undergoes a change inside the quantum well, can help improve device properties in cases where large polarization-induced built-in fields affect device operation.

Group III-Nitride Semiconductor Optoelectronics, First Edition. C. Jayant Praharaj.

Perpendicular Quantization and Density of States

The effective mass theorem is applicable for potentials that are slowly varying on the scale of the lattice constant. While quantum wells involve some fairly abrupt changes in the potential profile, calculations using the effective mass theorem have provided good agreement with experiments in a variety of material systems.

The effective mass time-independent Schrodinger equation for electrons in the conduction band is

$$-\frac{\hbar^2}{2m_\perp^*}\nabla^2\Psi+V\Psi=E\Psi \tag{4.1}$$

where ħ is the reduced Planck's constant (Planck's constant divided by 2Π), $m_\perp^*$ is the electron effective mass perpendicular to the quantum well, and *V* is the slowly varying potential above the crystal Hamiltonian. The eigenvalues of this equation are the allowed energy levels for perpendicular quantization. In the transverse direction, the usual E-k_x,k_y dispersion relation holds. The eigenfunctions that correspond to these eigenvalues are the envelope functions that need to be multiplied with the cell-periodic part to give the total wave function. While the effective mass theorem applies for slowly varying potential, good agreement with experiment has been observed for a wide class of materials in the case of abrupt heterojunctions like quantum wells.

The eigenfunctions corresponding to these eigenvalues are the envelope functions that need to be multiplied with the cell-periodic part to give the total wave function.

$$\psi_c(\mathbf{r})=u_c(\mathbf{r})\frac{exp(i\mathbf{k}_t.\mathbf{r}_t)}{\sqrt{A}}f(z) \tag{4.2}$$

The valence band is more complicated due to the presence of three kinds of holes: i) heavy, ii) light, and iii) spin-orbit split-off holes. We will discuss the valence-band case in more detail later in the chapter. Focusing on one type of holes, for example, the heavy holes, we have a similar kind of effective mass Schrodinger equation, and we get for the valence-band wave function in the quantum well

$$\psi_v(\mathbf{r})=u_v(\mathbf{r})\frac{exp(i\mathbf{k}_t'.\mathbf{r}_t)}{\sqrt{A}}g(z) \tag{4.3}$$

Figures 4.1 and 4.2 show typical conduction-band wavefunctions in the case of quantum wells without and with built-in electric fields. Figure 4.3 shows typical conduction-band offset at group III-Nitride heterojunctions.

The 3D DOS per unit volume

$$g(E)=\frac{1}{2\Pi^2}\left(\frac{2m^*}{\hbar^2}\right)^{\frac{3}{2}}\sqrt{E-Eco} \tag{4.4}$$

where *Eco* is the conduction-band edge energy and m^* is the effective mass that gets modified to the 2D DOS per unit area

$$g(E)=\sum_i\frac{1}{\Pi}\frac{2m_t^*}{\hbar^2}\,u(E-Eci) \tag{4.5}$$

where m_t^* is the transverse effective mass, *u* is the unit step function, and *Eci* is the *i*th energy level for perpendicular quantization as shown in Figure 4.4. The 2D DOS is lower than the bulk DOS.

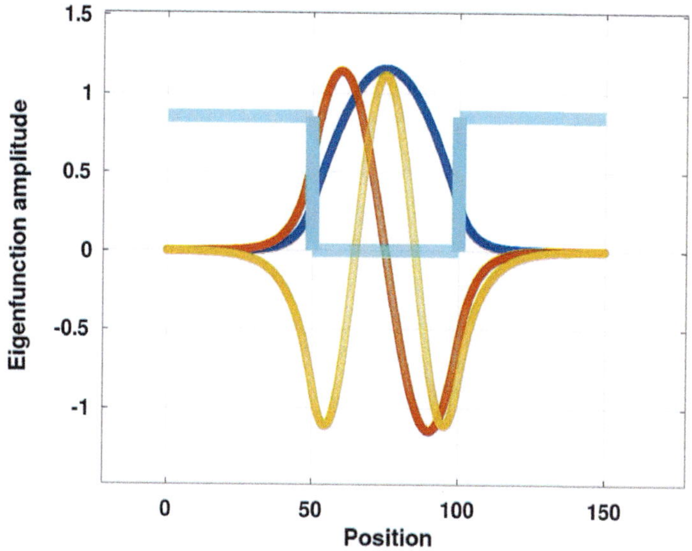

Figure 4.1 Conduction-band wave functions for a quantum well without a built-in electric field (*c*-axis wurtzite quantum wells have built-in electric fields and are shown in Figure 4.2).

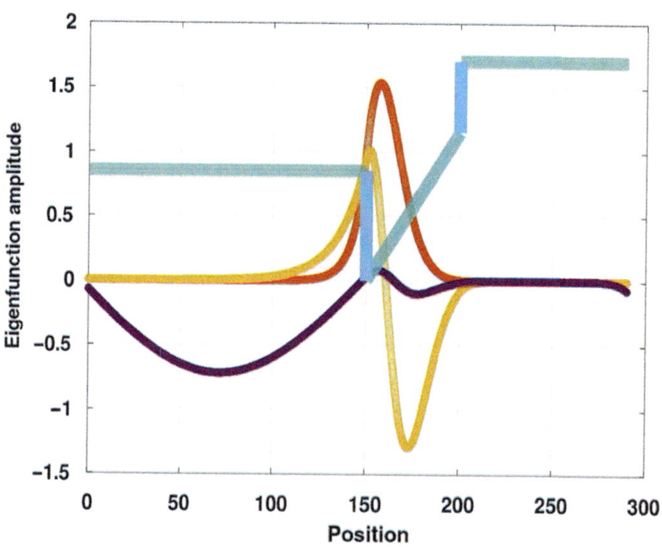

Figure 4.2 Conduction-band wave functions for a wurtzite quantum well with built-in electric field.

This means that the electron Fermi level must lie higher in the conduction band compared to the bulk case for the same effective electron concentration. It must be kept in mind that the three valence bands – the heavy hole, the light hole, and the spin-orbit split-off bands – have different densities of states. Figure 4.4 shows a 2D DOS for a quantum well.

For holes in the valence band, each type of hole (heavy hole, light hole, and the spin-orbit coupled split-off holes) has its own effective mass Schrodinger equation with the added complication

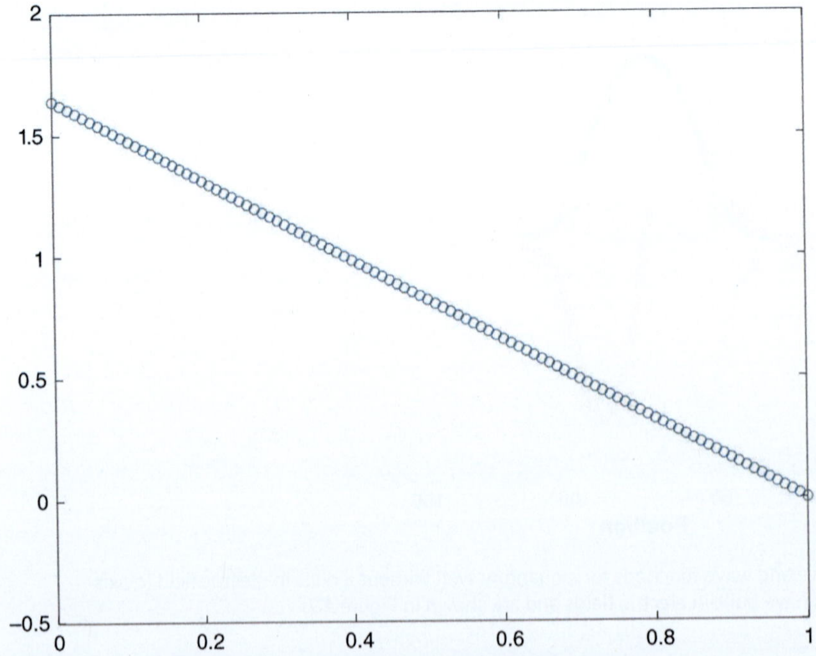

Figure 4.3 Conduction-band offset at the GaN/InGaN heterojunction.

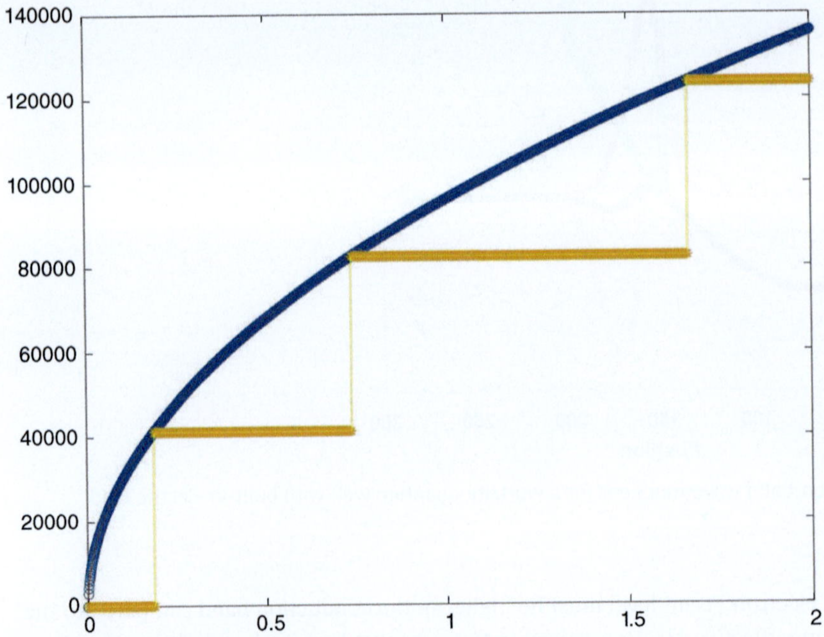

Figure 4.4 Modification of density of states (DOSs) from bulk to a 2D quantum well.

that there is coupling between the three types of holes. Numerical techniques become necessary to handle almost all valence-band hole problems in quantum wells.

The k.p Method

The k.p method expresses the time-independent Schrodinger equation for the Bloch wave function

$$H_0 \psi_{nk}(\boldsymbol{r}) = E_{\mathrm{nk}} \psi_{nk}(\boldsymbol{r}) \tag{4.6}$$

where H_0 is the crystal Hamiltonian, in terms of the cell-periodic part of the wave function, which becomes

$$H u_{nk}(\boldsymbol{r}) = \left(H_0 + \frac{\hbar^2 k^2}{2m_0} + \frac{\hbar}{m_0} \boldsymbol{k}.\boldsymbol{p} + H_{\mathrm{so}} \right) \psi_{nk}(\boldsymbol{r}) = E_{nk} u_{nk}(\boldsymbol{r}) \tag{4.7}$$

where H_{so} is the spin-orbit interaction term

$$H_{so} = \frac{\hbar^2}{4m_0^2 c^2} \nabla V \times \boldsymbol{p}.\boldsymbol{\sigma} \tag{4.8}$$

Perturbation theory can be used to solve this equation by treating the terms in the Hamiltonian other than H_0 as a perturbation. Diagonalization can also be used in specific bases that are suitable for the problem at hand. The eight-band k.p Hamiltonian within the Kane model can be used to obtain dispersion relations and quantized energies. The basis chosen for writing the Hamiltonian is as follows

$$u_1 = \left| \frac{X + iY}{\sqrt{2}} \uparrow \right\rangle,\ u_2 = \left| \frac{X - iY}{\sqrt{2}} \uparrow \right\rangle, u_3 = \left| Z \uparrow \right\rangle,\ u_4 = \left| \frac{X + iY}{\sqrt{2}} \downarrow \right\rangle,\ u_5 = \left| \frac{X - iY}{\sqrt{2}} \downarrow \right\rangle, u_6 = \left| Z \downarrow \right\rangle,\ u_7 = \left| S \uparrow \right\rangle,\ u_8 = \left| S \downarrow \right\rangle$$

$$H_{8\times8} = \frac{\hbar^2 k^2}{2m_0} + \begin{bmatrix} \mathrm{Ec} & \frac{-\mathbf{k_+ P_2}}{\sqrt{2}} & \frac{\mathbf{k_- P_2}}{\sqrt{2}} & \mathrm{k_Z P_1} & 0 & 0 & 0 & 0 \\ -\frac{\mathbf{k_- P_2}}{\sqrt{2}} & \mathrm{Ev}+\Delta_1+\Delta_2 & 0 & 0 & 0 & 0 & 0 & 0 \\ \frac{\mathbf{k_+ P_2}}{\sqrt{2}} & 0 & \mathrm{Ev}+\Delta_1-\Delta_2 & 0 & 0 & 0 & 0 & \sqrt{2}\Delta_3 \\ \mathrm{k_Z P_1} & 0 & 0 & \mathrm{Ev} & 0 & 0 & \sqrt{2}\Delta_3 & 0 \\ 0 & 0 & 0 & 0 & \mathrm{E_C} & \frac{\mathbf{k_- P_2}}{\sqrt{2}} & \frac{-\mathbf{k_+ P_2}}{\sqrt{2}} & \mathrm{k_Z P_1} \\ 0 & 0 & 0 & 0 & \frac{\mathbf{k_+ P_2}}{\sqrt{2}} & \mathrm{Ev}+\Delta_1+\Delta_2 & 0 & 0 \\ 0 & 0 & 0 & \sqrt{2}\Delta_3 & -\frac{\mathbf{k_- P_2}}{\sqrt{2}} & 0 & \mathrm{Ev}+\Delta_1-\Delta_2 & 0 \\ 0 & 0 & \sqrt{2}\Delta_3 & 0 & \mathrm{k_Z P_1} & 0 & 0 & \mathrm{EV} \end{bmatrix}$$

where the various elements of the 8×8 matrix are composed of interband momentum matrix elements and interband crystal Hamiltonian matrix elements. The effective mass approximation now involves both first and second order derivatives of the wave function as well as the dispersion of some of the bands that involve terms from other bands.

Since the conduction-band and valance-band states are separated by a large bandgap, we can split the eight-band problem into a two-band conduction-band problem and a six-band valence-band problem. A more accurate method for compound semiconductor problems is the Luttinger–Kohn method. In this, the six valence bands that are close by in energy are treated as one group, and the two conduction bands are treated as another group.

The Hamiltonian for the valence-band part can be expressed as the sum of the Hamiltonian for the unstrained semiconductor plus a Hamiltonian for the strain part. The Hamiltonian for the unstrained part is written as the sum of a $k=0$ part plus a k-dependent part H_1.

Where

$$F=\Delta_1+\Delta_2+\lambda+\theta \tag{4.9}$$

$$G=\Delta_1-\Delta_2+\lambda+\theta$$

$$\lambda=\frac{\hbar^2}{2m_0}\left[A_1k_z^2+A_2\left(k_x^2+k_x^2\right)\right]$$

$$\theta=\frac{\hbar^2}{2m_0}\left[A_3k_z^2+A_4\left(k_x^2+k_x^2\right)\right]$$

$$K=\frac{\hbar^2}{2m_0}\left(k_x+ik_y\right)^2$$

$$H=\frac{\hbar^2}{2m_0}\left(k_x+ik_y\right)k_z$$

$$\Delta=\sqrt{2}\Delta_3$$

A_1, A_2, A_3, A_4, etc. are parameters of the theory that are defined in terms of momentum matrix elements, bandgap, etc. When strain is present in the material, the parameters are modified as follows

$$\lambda=\frac{\hbar^2}{2m_0}\left[A_1k_z^2+A_2\left(k_x^2+k_x^2\right)\right]+\lambda_\varepsilon \tag{4.10}$$

where

$$\lambda_\varepsilon=D_1\varepsilon_{zz}+D_2\left(\varepsilon_{xx}+\varepsilon_{yy}\right)$$

$$\theta=\frac{\hbar^2}{2m_0}\left[A_3k_z^2+A_4\left(k_x^2+k_x^2\right)\right]+\theta_\varepsilon$$

$$\theta_\varepsilon=D_3\varepsilon_{zz}+D_4(\varepsilon_{xx}+\varepsilon_{yy})$$

The parameters in these matrixes can be computed using ab initio density functional techniques or using semi-empirical tight-binding methods. In a quantum well perpendicular to the z direction, the k_z variable becomes the operator $-i\hbar\frac{\partial}{\partial z}$ for the evaluation of sub-band dispersion, etc (the multi-band analog of the single-band effective mass approximation). Similarly, for a quantum wire, we get $-i\hbar\frac{\partial}{\partial z}$ and $-i\hbar\frac{\partial}{\partial x}$ when the wire runs along the y direction.

The gain or loss (absorption) coefficient is given by

$$g(\hbar\omega)=\frac{2q^2 \Pi}{nc\varepsilon_0 m_0^2 \omega}\sum\nolimits_{n,m}\left|I_{\rm hm}^{\rm en}\right|^2 \frac{2}{V}\sum\nolimits_{k_t}\left|\boldsymbol{e}.\boldsymbol{p}_{\rm cv}\right|^2 \delta\left(E_g+E_{\rm en}-E_{\rm hm}+\frac{\hbar^2 k_t^2}{2m_r}-\hbar\omega\right)\left(f_v^m-f_c^n\right) \quad (4.11)$$

where q is the magnitude of the electron charge, m_0 is the electron mass in free space, and c and ε_0 are the velocity of light and permittivity in free space, respectively. $\boldsymbol{e}$ is the polarization vector of the optical electric field; n and Lw are the refractive index and thickness of the quantum well (QW), respectively. In the momentum matrix element, the cell-periodic part momentum matrix element is multiplied with the overlap integral between conduction and valence-band states to a high level of accuracy.

Characterization of Quantum Wells

Photoluminescence is one of most commonly used methods to understand the carrier quantum states and their correlation with optical transitions. While optoelectronic devices can have extra complications like free-carrier screening, these studies of quantum-well layers provide good

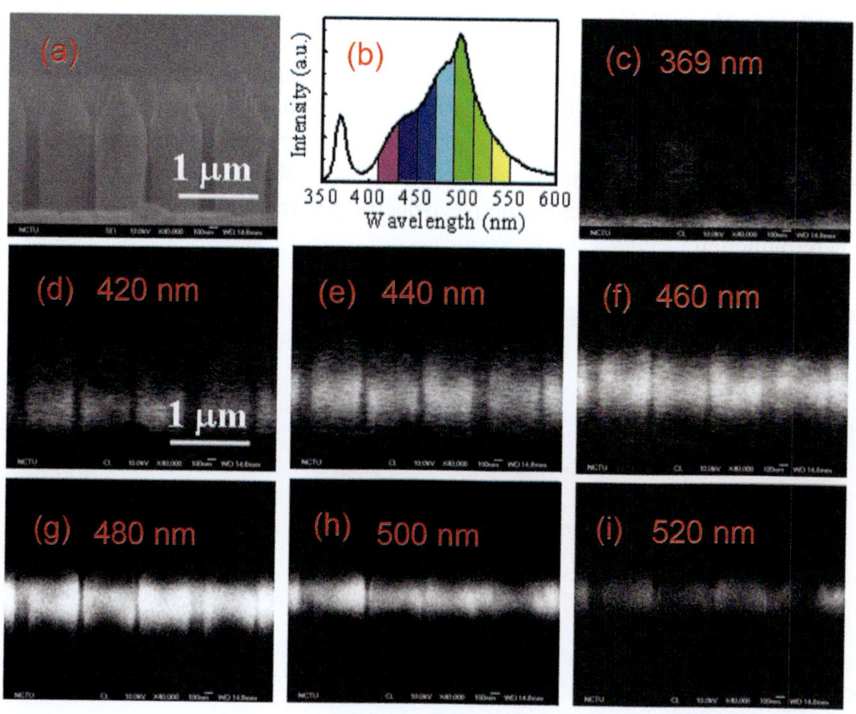

Figure 4.5 Cathodoluminescence (CL) spectra of nanopillar quantum wells (from Reference [2]).

insight into the conduction and valence-band states as well as the kind of emission that can be expected in devices. Furthermore, photoluminescence can typically provide a good understanding of deep states that can act as non-radiative recombination centers. Cathodoluminescence is another method of characterization that yields knowledge about structure and spectra. Figure 4.5 shows typical cathodoluminescence spectra in nitride quantum wells.

Spontaneous and Piezoelectric Polarization Effects

Wurtzite quantum wells have built-in polarization discontinuities of the spontaneous and piezoelectric variety. This can lead to significant modification of optical emission properties and can sometimes be a limiter for optical device performance. The magnitude of spontaneous polarization along the *c*-axis is given by

$$P_{sp}\left(AlxGa1-xN\right)=-0.090x-0.034\left(1-x\right)+0.021x\left(1-x\right)\ Cm^{-2}\left(1\right) \tag{4.12}$$

$$P_{sp}\left(InxGa1-xN\right)=-0.042x-0.034\left(1-x\right)+0.037x\left(1-x\right)Cm^{-2}\left(2\right)$$

$$Psp\left(AlxIn1-xN\right)=-0.090x-0.042\left(1-x\right)+0.070x\left(1-x\right)Cm^{-2}\left(3\right)$$

where the direction of polarization is along (0001).

The piezoelectric polarization along (0001) is given by

$$P_{piezo}=2\left(\left(a-a_0\right)/a_0\right)\left(e_{31}-e_{33}C_{13}/C_{33}\right)\ Cm^{-2}\left(2\right) \tag{4.13}$$

where a_0 is the equilibrium lattice constant in the hexagonal plane, and e_{31}, e_{33}, C_{13}, C_{33} are piezoelectric and elastic constants, respectively. Here, the elastic constants can be taken as linear interpolation in the alloys as a first approximation. However, accurate determination of piezoelectric polarization requires numerical evaluation using advanced techniques like density functional calculations. Figure 4.6 shows typical spontaneous and piezoelectric polarization configurations at a group III-Nitride heterojunction.

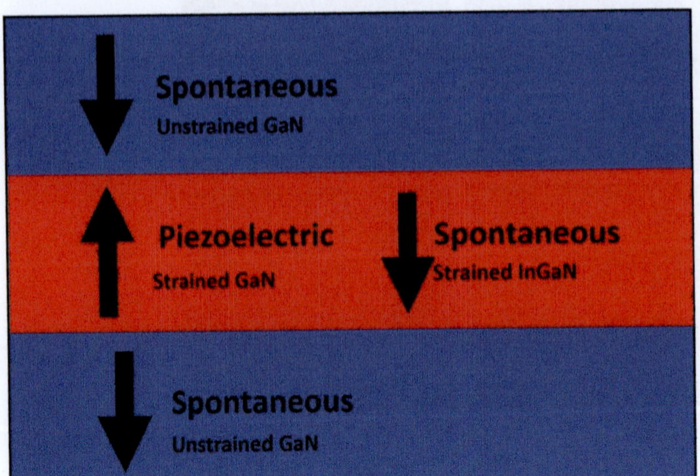

Figure 4.6 Directions of spontaneous and piezoelectric polarization at a typical group III-Nitride quantum well.

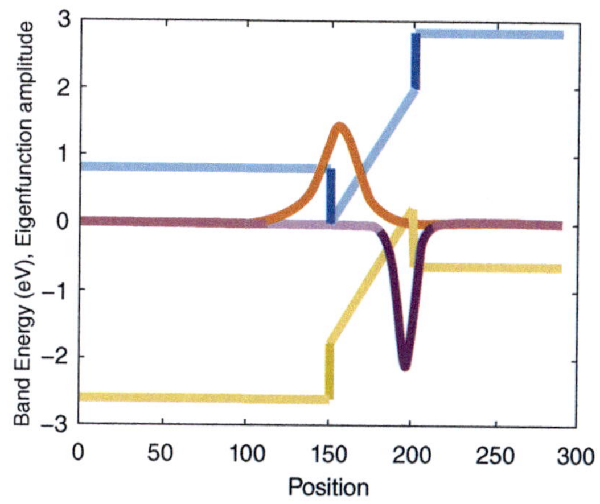

Figure 4.7 Reduced electron and hole wave-function overlap due to polarization-induced built-in electric fields.

Spontaneous and piezoelectric polarization discontinuities in wurtzite nitride quantum wells lead to interface bound charges on the order of 10^{13} electrons per cm^2 and they can be of either sign depending on the specific well structure being used. Basic electrostatic considerations lead to electric field discontinuities of several mega volts (MV) per cm and large magnitudes of band bending. The most significant physical consequences for optical transitions are i) severe red shift of the optical frequency corresponding to energy difference between quantum-well states and ii) suppression of emission efficiency due to poor overlap between conduction-band and valence-band quantized state wave functions (as shown in Figure 4.7).

As will be discussed in Chapter 5 on optoelectronic devices that uses these quantum wells, screening by free charge carriers (electrons and holes) can lead to some improvement in these characteristics when optical transitions are set in motion by current flow due to applied bias, but significant barriers to efficient optical emission remain.

In the wurtzite material system, *m*-plane and *a*-plane growth can lead to polarization-free quantum wells and a better overlap of conduction- and valence-band wave functions. The modification of dislocation densities can lead to differences in device performance in device structures. As discussed in Chapter 3, *m*-plane growth can occur on *r*-plane sapphire substrate.

Staggered well designs can improve the conduction/valence-band overlaps by modifying the overall amount of band bending in the quantum well. However, the device design and the fabrication of devices becomes that much more complicated.

Quantum Dots and Quantum Wires

The Stranski–Krastanow (SK) growth mode in molecular beam epitaxy (MBE) has become the most common method of obtaining QDs in the group III-Nitride materials system. InGaN quantum dots embedded in wider bandgap material like GaN are of great technological interest and comprise the most widely fabricated and studied QD system in group III nitrides. SK growth leads to pyramidal InGaN QDs embedded in GaN or similar material.

For quantum wires, the DOS gets modified to a 1D-DOS given by Equation 4.14, and for QDs, the DOS is given by a sum of delta functions as shown in Equation 4.15.

$$g(E) = \sum_i \frac{1}{\Pi} \frac{C}{\sqrt{E - E_{ci}}} u(E - E_{ci}) \tag{4.14}$$

$$g(E)=\sum_i \delta(E-E_i) \tag{4.15}$$

The k.p calculations for QDs can yield the energy spectrum of electrons and holes. Figures 4.8 and 4.9 show scanning electron microscopy (SEM) and transmission electron microscopy (TEM) images of typical nitride QDs, respectively. Figure 4.10 shows the calculated spectrum for typical InGaN QDs using an eight-band k.p model. Figure 4.11 shows the emission-based characterization of typical InGaN QDs.

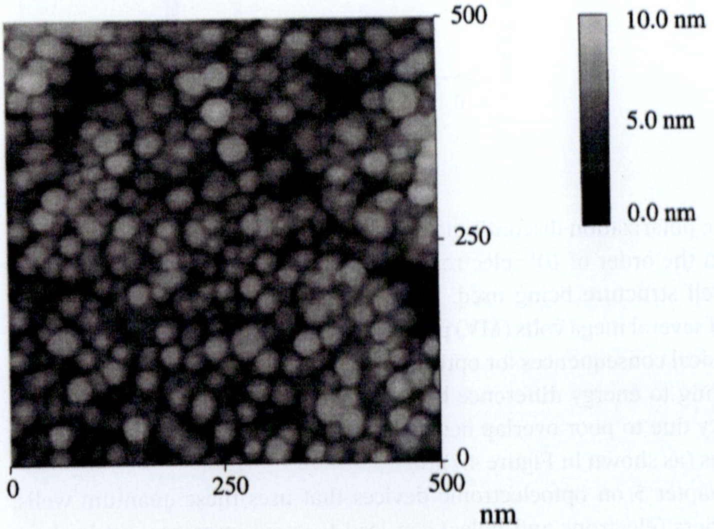

Figure 4.8 Self-assembled $In_{0.35}Ga_{0.65}N$ QDs grown on GaN (Reference [3]).

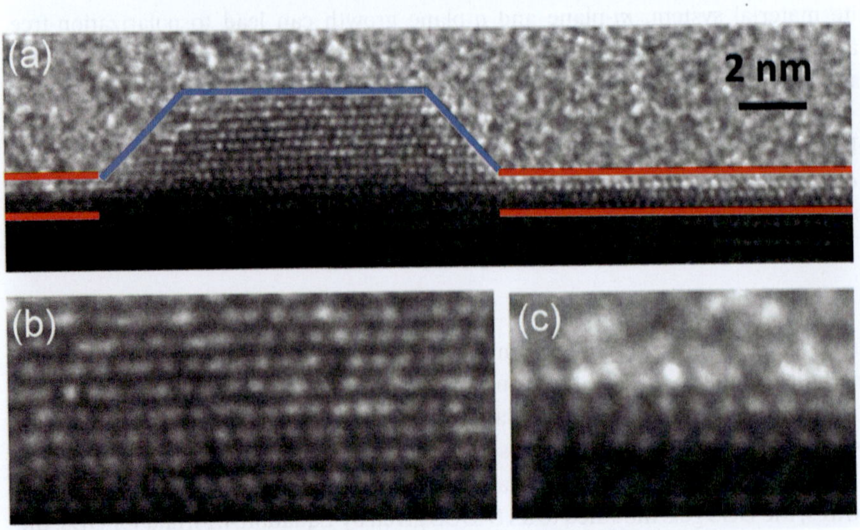

Figure 4.9 High-resolution transmission electron microscopy (HRTEM) images of indium nitride (InN)/$In_{0.73}Ga_{0.27}N$ Stranski–Krastanow QDs grown on (111) silicon (Si) (Reference [4]).

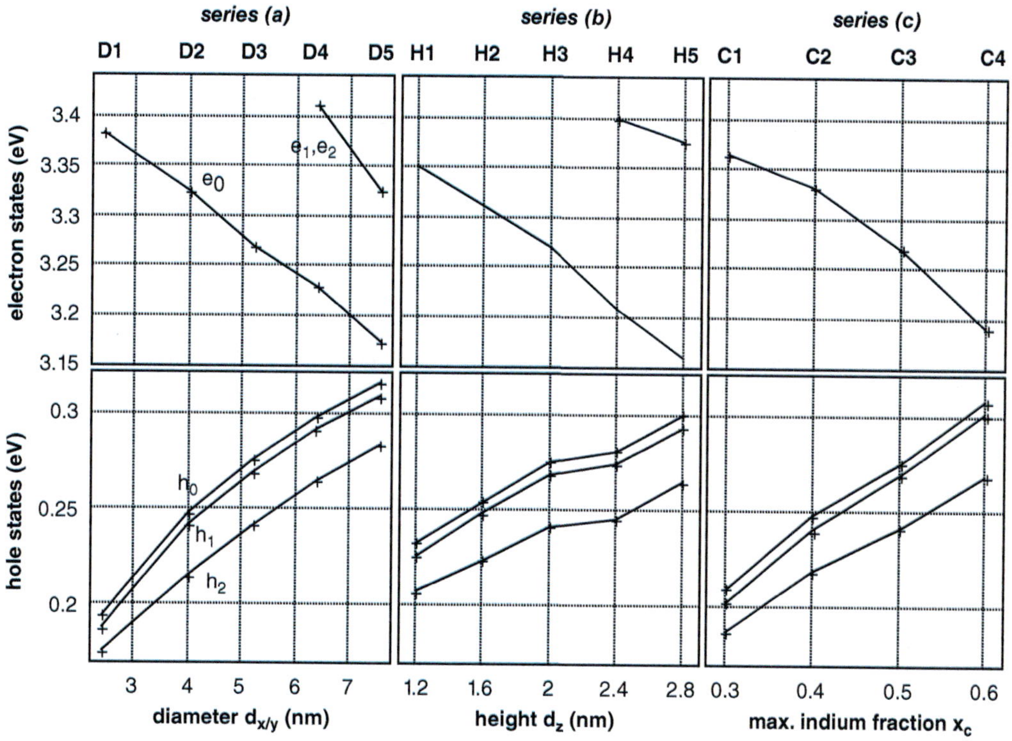

Figure 4.10 Electron and hole energies in ellipsoidal InGaN QDs: varying diameter (series [a]), varying height (series [b]), and varying In concentration (series [c]) (after Reference [5]).

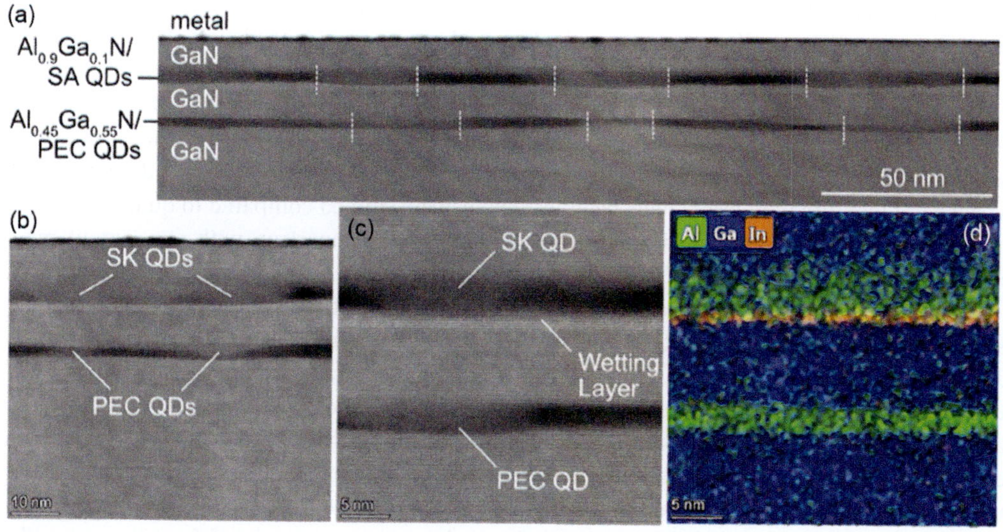

Figure 4.11 HRTEM image of Stranski–Krastanov InGaN QDs with AlGaN layer and GaN barrier for near-infrared emission. From Reference [6], used with permission.

Crystal Symmetry and Low-dimensional Structures

The wurtzite crystal symmetry of group III-Nitride semiconductors, and in particular InGaN quantum-well materials, manifests itself in quantum mechanical states, their energy levels and energy dispersions (E versus k), and the occupation of these energy levels. This is also true for low-dimensional structures in many cases. However, there are subtleties in how much the properties of QWs, quantum wires, and QDs are affected by the underlying wurtzite crystal symmetry.

First, in QWs the crystal symmetry is explicitly broken in the direction perpendicular to the QW. Similarly, for quantum wires the crystal symmetry is broken in the two directions perpendicular to the quantum wire axis. It should also be kept in mind that in a quantum well like InGaN/GaN, for example, the crystal parameters are different between the well layer (InGaN) and the barrier layer (GaN) even in the 2D plane of the quantum well. In this specific example, the pseudopotentials of the InGaN layer are different from the pseudopotentials in the GaN layer. The reader is urged to refer to books on electronic structure calculation for more details on pseudopotentials and the role they play in deciding the electronic structure and E-k dispersions of semiconductors. The fact that the pseudopotentials for the InGaN alloy are different from that of the GaN barrier layer means that careful boundary condition matching may be needed to obtain accurate energy levels of quantum wells and their dispersion. The effective mass approximation and the k.p method actually give good agreement between theory and experiments for structures like InGaN/GaN quantum-well lasers and LEDs. However, research in this field has extended to extreme structure like delta quantum wells where the quantum-well layer is only a few atomic layers thick and not 3–5 nm, as in typical quantum wells. There is no guarantee that the k.p method will continue to give accurate energy dispersions and energy levels when these more extreme quantum-well designs are used, for example in ultraviolet (UV) emitters.

The case of quantum wires is even more complicated because the quantum wire can have different cladding on its four sides. For example, one can have a case where the barrier layer is of one type on three sides of the quantum wire and of another type on the fourth side of the quantum wire. This means that pseudopotentials inside the wire layer and inside the barrier layers can be of three or more types. This means that, to be accurate, the crystal symmetry is broken even in the direction parallel to the quantum wire (the wire layer and barrier layers having different microscopic potential profiles) and problems with matching wave functions in the direction parallel to the quantum wire can come up. This is true even when the quantum wires are pseudomorphic to the barrier. Quantum wire lasers have been less extensively studied compared to quantum wells, and the accurate determination of quantum wire states for comparison with experiment may require one to go beyond simple k.p methods.

QDs probably provide the most extreme cases of crystal symmetry breaking. When one tries to use k.p methods to find QD states it must be kept in mind that writing the total potential in terms of barrier potential for applying the k.p method to QD electronic structure is extremely complex. Like quantum wires, the barrier layers can be different on different faces of the QD. Also, the QD may not have a simple cuboidal shape, and it may have pyramidal shape. This means crystal symmetry can become much less relevant to the electronic structure, especially for small dots. Fully crystal symmetry-broken methods may be required to calculate QD energies for advanced QD structures.

Another complication from the perspective of crystal symmetry can arise if the quantum wells or quantum wires have well-width or wire-width fluctuations. Epitaxial methods like MBE can be optimized to give extremely planar growth. However, in any particular growth method, the possibility of width fluctuations at the atomic scale cannot be completely ignored. Width fluctuations make the boundary conditions for matching within-well (or within-wire) with the within-barrier parts of the wave function non-trivial. For advanced structures, the effect can be significant enough that simple k.p-type effective mass methods may lead to erroneous conclusions.

The deviations from crystal symmetry and symmetry-based calculation methods become more and more serious as advanced structures are explored. For example, for a pyramidal QD, the wurtzite crystal symmetry is broken in all directions even inside the QD. Further, applying boundary conditions to pyramidal QDs is extremely difficult. Also, if dimension fluctuations are added to the pyramidal shape, we are left with a highly non-trivial electronic structure problem. One of the ways is to use highly complex calculation methods like supercell-based density functional theory (DFT) calculations, to be discussed in the next section.

Another consequence of reduction in crystal symmetry is that many of the simple methods used to calculate phonon dispersion relations (lattice vibration energy spectra) for heterostructures can become insufficient. Again, using the example of pyramidal QDs, the boundary conditions needed to evaluate the phonon spectra in the QDs becomes extremely complex. The reader must remember that high levels of crystal symmetry lead to drastic simplifications in electronic energy spectra and phonon spectra. Deviations from crystal symmetry likewise make these spectra more complex. Once again, supercell-based DFT calculations may become necessary. Remember that accurate knowledge of phonon spectra can become relevant in many optoelectronic devices because they play a role in determining scattering rates and relaxation rates, which in turn affect device speed. Going from QDs to quantum wires again many of the symmetry-based simplifications for bulk semiconductors become unavailable, which leads to the requirement for sophisticated calculation and huge computational resources.

Raman scattering experiments can shed some light on optical phonon modes and can be used as a guide to theoretical methods. However, they do not give the full dispersion, and it is a priori not clear how much dispersion happens in advanced quantum wires and QDs. Neutron scattering methods are difficult to interpret for quantum wells, quantum wires, and QDs because the phononic contribution from the well, wire, or dot can be mixed with the bulk phononic contribution. Also, acoustic phonon branches can become very complex due to deviations from crystal symmetry, and the acoustic phonon scattering part of carrier dynamics becomes that much more difficult to understand.

Another aspect of carrier dynamics that becomes more complicated with the reduction in symmetry is plasmon collective oscillations of electron gases and hole gases in device structures with wells, wires, and dots. Plasmon calculations based on simple effective mass methods are likely to be inaccurate the more the deviation from crystal symmetry there is. That means that understanding collective oscillations like plasmons in advanced heterostructure devices can require advanced atomistic calculations. The problem is even more serious for valence-band holes compared to conduction-band electrons because there are three kinds of holes – heavy, light, and spin-orbit split-off holes. The collective oscillations from these kinds of holes can intersperse and lead to complicated plasmon spectra. Plasmons, like phonons, affect carrier scattering rates, especially when electron densities and hole densities are high. So, a reduction in crystal symmetry makes it that much more difficult to understand and quantify scattering rates as well as the effect of scattering rates on device speeds.

Alloy Disorder and Density Functional Theory Electronic Structure Calculation

InGaN and aluminum gallium nitride (AlGaN) ternary quantum wells have alloy disorder, which can lead to some unique problems for understanding the electronic structure and quantum mechanical energy levels in these quantum wells. This is also true for quaternary alloys like InAlGaN,

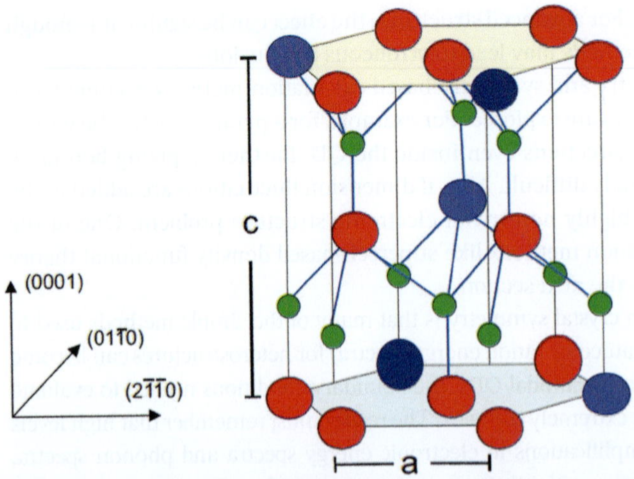

Figure 4.12 Level of alloy disorder can lead to deviations from k.p theory.

which is of potential interest for some efficiency improvement designs in lasers and LEDs. Figure 4.12 shows a schematic diagram of the kind of alloy disorder seen in group III-Nitride alloys.

Low levels of alloy disorder can lead to regimes where the k.p parameters for the InGaN alloy is an average of the parameters for InN and GaN. Similar conclusions hold when the level of alloy disorder is low in AlGaN. However, for high levels of alloy disorder in InGaN and AlGaN, simple k.p theory modifications can lead to erroneous conclusions regarding quantum mechanical states, scattering rates, conductivity, etc.

For high levels of disorder, one way of arriving at accurate electronic structure, electronic spectra, and energies is to use a first-principle electronic structure calculation method like DFT. This is not a book on electronic structure calculations, so the reader is advised to consult one of several excellent books available on the subject of DFT electronic structure calculation of semiconductors. For this book, we are interested in how DFT or similar methods can help us understand InGaN quantum wells, quantum wires, and QDs accurately when the alloy disorder is high.

To calculate the electronic structure for an InGaN alloy QD, for example, one can define a supercell. The supercell is basically a structure where the QD and its barrier layers are together considered as one unit cell. The supercell obviously contains a large number of atoms. So, the computational resources necessary to calculate electronic structure and spectra become large. One of the conclusions for large levels of alloying is that electron and hole states may become more highly localized compared to states calculated using k.p methods.

For quantum wells and quantum wires, again, localization of electronic states due to alloy disorder can become significant enough that k.p methods give the wrong conclusions. While using supercell methods for QDs can be achieved using powerful computational resources and supercomputers, doing the same kind of calculation for quantum wells and quantum wires can be prohibitive even on supercomputers.

Alloying, due to the random potential it introduces, can lead to significant localization of electron and hole states and deviations from the Bloch theory of electronic states. The effect on optoelectronic device operation is also non-trivial. The localization of states can lead to large modifications of quantities like emission rates and quantum efficiencies of devices (to be discussed later in this chapter).

The combination of high levels of alloying and large deviations of crystal symmetry becomes especially relevant for many scaled optoelectronic devices with wells, wires, and dots. This basically superimposes one kind of symmetry reduction on another kind of symmetry reduction. The need for atomistic calculations becomes all the more urgent.

Another problem associated with high levels of alloying is the fact that the phonon spectra can become very complicated. The spring constants associated with indium-nitrogen (In–N) bonds and those associated with Ga–N bonds are different. Also, they vary with the level of alloying. This means that phonon spectra can have very non-trivial dependences on the level of alloying. Understanding phonon spectra in alloyed heterostructures is key to understanding phonon scattering and its effect on device speed and power performance for advanced structures.

The next problem that needs to be understood when the level of alloying is high is the effect on collective oscillations like plasmons of the randomness in the crystal potential. Simple extrapolations of effective mass from compounds to alloys and simple extrapolations of k.p parameters can lead to mistakes in quantitative predictions. Atomistic calculations of plasmon spectra and their effect on carrier scattering and device parameters become necessary in many cases.

Phonon Scattering in Quantum Wires and Quantum Dots

The modification of the electron DOS in quantum wires and QDs has significant consequences for electron–phonon scattering in these structures. Since in the perturbation theoretic limit the scattering rates are strongly dependent on final DOS, severe suppression of electron–phonon scattering rates takes place. For example, in QDs, a relatively high energy phonon of the order of 100 meV is needed to cause a transition from one QD state to another. If this phonon is from the acoustic branch, a large *k*-vector on the order of $10^9\ m^{-1}$ is associated with it. That means that the scattering matrix element associated with it depends strongly on the spatial variation of the QD initial and final state wavefunctions. If these wave functions have spatial frequency variation on the order of the phonon *k*-vector required, then good scattering amplitudes can be obtained. With a 5 nm QD, the spatial frequency is typically less than $5\times10^8\ m^{-1}$. This can seriously hamper the efficacy of scattering. Optical phonons can provide a large energy with low *k*-vector. However, the dispersions of the optical branches tend to be low, and energy matching with inter sub-band levels is difficult to achieve. This is the so-called “phonon blockade” problem in QD LEDs.

The problem is likely to be more serious for a moderately sized QD than for an aggressively scaled QD. However, Stranski–Krastanow-type growth often leads to QD sizes in a range and patterned QDs require expensive fabrication techniques like electron-beam lithography, and they have the further problem of interface roughness. Also, aggressively scaled QDs have the added problem that the phonon spectra can become drastically modified from bulk behavior, making them that much more difficult to quantify.

Quantum wires tend to have lower scattering rates from phonons compared to QWs due to the lower carrier DOS discussed earlier. Whether or not this affects the speed of quantum wire lasers depends on whether these scattering rates are lower than the speed limits discussed in Chapter 5, that is whether these scattering rates provide the bottleneck for speed compared to other device parameters.

Plasmons in Quantum Wells and Quantum Wires

Under high bias in LEDs and lasers, the conduction-band 2D electron gas (2DEG) density and the valence-band hole gas (2DHG) density can become very large, with values possible in the 10^{13} per

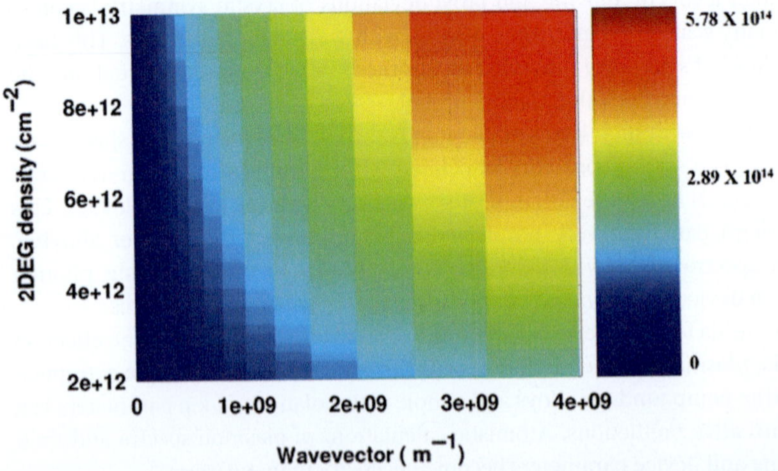

Figure 4.13 2DEG plasmon frequencies in sec^{-1} for a typical 2DEG density range.

cm^2 range. This leads to large plasmon frequencies so that dissipative damping of plasmons cannot damp out the plasmons completely. The plasmon frequencies can get close to optical phonon frequencies so that they can hybridize with and couple to the plasmons to form coupled phonon–plasmon modes. The scattering rates of electrons and holes may need to be calculated based on these coupled modes, depending on the level of damping of the plasmon modes. Coupled modes can lead to more efficient scattering and energy relaxation for both electrons and holes injected from the barrier layer into the QW. Figure 4.13 shows wavevector-dependent plasmon frequencies in two-dimensional electron gases (2DEGs).

The plasmons associated with valence-band holes are more complex than those for electrons. The heavy holes, light holes, and spin-orbit split-off holes can lead to complex plasmon dynamics and the problem becomes even more complex due to well-width fluctuations, wire-width fluctuations, and QD shapes. The problem becomes especially severe for highly scaled structures like delta quantum-well LEDs and lasers, which have well layers only two or three atomic layers thick.

Deviations from Charge Neutrality and Effect on Electronic Structure

As discussed previously, the large polarization-induced built-in electric fields in QWs, quantum wires, and QDs grown along the *c*-axis or (0001) direction lead to significant spatial separation of electron wave functions and hole wave functions. This is the intrinsic quantum-confined Stark effect (QCSE) in these structures. This means that the electrons tend to accumulate on one side of a QW and holes on the other side. This leads to deviations from charge neutrality on the two sides of the QW. Deviations from charge neutrality compared to the bulk means that the k.p parameters in the well can differ significantly compared to bulk values. For certain combinations of well layers and barrier layers, the change in k.p parameters can lead to quantitatively significant changes in electronic structure parameters like sub-band energies, transverse effective masses, and transverse energy dispersion.

It must be kept in mind that the same kind of separation between electrons and holes happens for quantum wires and QDs. The question of electronic structure in quantum wires and QDs is understood even less compared to QWs. For pyramidal QDs, the electron and hole wave-function separation leads to inhomogeneous electron liquids and inhomogeneous hole liquids. For homogeneous electron gases and hole gases, derived expressions are available to account for direct Coulomb energy and exchange-correlation energy, which can be used to improve sub-band energy and electronic structure calculations. However, inhomogeneous electron liquids and hole liquids constitute a much more difficult system to understand electronic structure. The calculation of energies in these QDs for comparison with experiment is a subject of ongoing research.

Polarization Engineering Using Quaternaries and Complex Structures

Overcoming Polarization-induced Red Shifts and Increasing Optical Matrix Element Magnitudes Using Quaternaries

Quaternary alloys provide an additional design paradigm, and some interesting material composition combinations have been found that can potentially lead to significant cancellation of spontaneous and piezoelectric polarization effects, which leads to flatter quantum-well profiles and corresponding improvement to the red-shift problem. In fact, significant progress can be made toward reducing polarization-induced charges and electric fields using ternary alloy compositions. However, devices often tend to be sensitive enough that more sophisticated solutions like quaternaries are needed. Specifically, AlInGaN can potentially be lattice matched to GaN or some InGaN composition, so that the piezoelectric component of polarization can be completely eliminated. Figure 4.14 shows lattice constant variations for InAlGaN quaternaries that can be used to obtain lattice matching.

Specific compositions of AlInGaN can be used to obtain polarization matching or close polarization matching with ternary group III-Nitride quantum-well layers. The optical emission efficiency

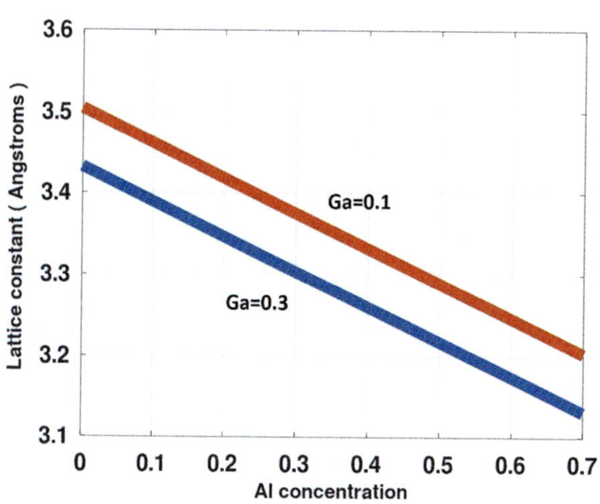

Figure 4.14 Lattice constant versus Al concentration of quaternary AlInGaN alloy for Ga concentrations 0.3 and 0.1.

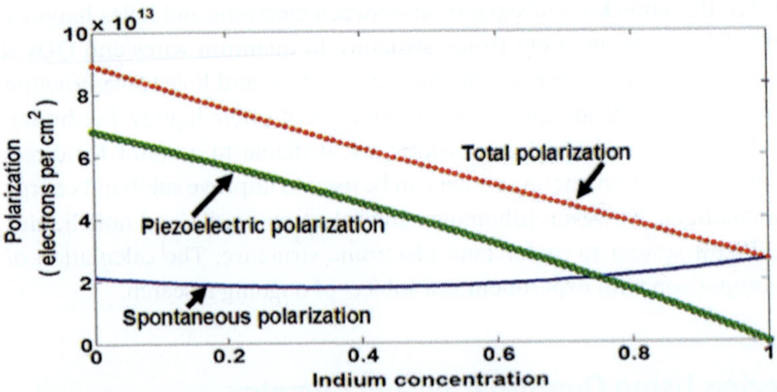

Figure 4.15 Polarization components for InGaN grown on GaN. The piezoelectric component is present only for pseudomorphic layers (layer thickness < critical thickness).

can therefore be increased compared to structures with built-in fields. Quaternary well layers or barrier layers are difficult to grow epitaxially, however. Therefore, device performance versus processing complexity trade-offs must be considered in any specific design.

The band bending due to the built-in polarization-induced fields can easily be greater than the conduction and valence-band discontinuities. This means that some of the lowest eigenstates (or highest for the valence band) can have significant carrier probability density in the barrier layer. Depending on the occupation of conduction and valence-band states, transitions from the barrier layers into the confined layers becomes possible. Non-polar wells grown on the *m*-plane, for example, will not have this kind of overlap with barrier layer states. Figure 4.15 shows typical polarization-induced charges of InGaN on GaN, which can be alleviated using quaternary technology.

Growth of quaternary alloys requires optimization of substrate-epitaxial layer interactions. Growth temperature and thermal mismatch of epitaxial layers can lead to novel regimes as far as dislocation densities and defect densities are concerned. Furthermore, the critical thicknesses can have complicated dependencies on substrate and growth conditions like temperature and pressure so that empirical methods become crucial to determining the onset of dislocation formation and the thickness limits of pseudomorphic layers.

Quasi-continuous Grading and Band Bending

Quasi-continuous grading has been used successfully in heterojunction bipolar transistors in some material systems to obtain better frequency responses. This kind of grading during MBE or metal-organic chemical vapor deposition (MOCVD) growth can lead to bulk polarization-induced bound charges that can help alleviate the band-bending problem. The polarization-induced charge is given by

$$\rho_b = -\nabla.\mathrm{P} \tag{4.16}$$

where ρ_b is the bound charge density and P is the polarization. Since the strain in InGaN layers tends to be high due to the large lattice mismatch between InN and GaN even moderate grading can lead to a lot of decrease in band bending when it is of the right sign. The quantum well, wire, or dot states are affected by the grading, but the barrier height can still be kept large enough to confine the electrons and holes effectively.

Dislocations in Low-dimensional Structures and Carrier Dynamics

Critical Thicknesses and Dislocation Formation

The lattice mismatch between InN and GaN is large, and even low indium content InGaN can have a large percentage lattice mismatch. The critical thickness for the formation of misfit dislocations as a function of indium concentration can be calculated according to two different models, the Matthews–Blakeslee and the Bean–People.

It has to be kept in mind that the actual critical thickness can depend on kinetic factors during non-equilbrium growth and often needs to be verified empirically. The actual critical thicknesses can be different for wells, wires, and dots due to kinetic considerations.

Recombination Dynamics of Carriers

A key feature of light-emitting as well as energy-conversion devices is the carrier recombination dynamics in the presence of dislocation-related or other type of traps. The overall recombination rate is given by

$$R = An + Bn^2 + Cn^3 \tag{4.17}$$

where the An term is due to Shockley–Read–Hall-type recombination via mid-gap states, the Bn^2 term is due to direct recombination of conduction-band electrons and valence-band holes, and the Cn^3 term is due to Auger recombination in which the electron recombines by giving energy to another conduction-band electron.

Disorder, Carrier Localization, and Effect on Recombination and Red Shifts

Ab Initio DFT Calculations

Since InGaN alloys have alloy disorder, the Bloch states of periodic crystals can become drastically modified. Most of the carrier effective mass determinations using techniques like electron cyclotron resonance and infrared ellipsometry have been done on non-alloyed GaN and InN. Theoretical work points to the bowing of InGaN effective mass as the composition is changed from GaN to InN.

Ab initio DFT calculations can often provide better insight into alloyed regimes compared to tight-binding or linear combination of atomic orbital (LCAO) calculations. Since there is significant alloy disorder in many of the InGaN layers used in devices, large unit cells need to be used to calculate energy spectra and E-k dispersion relations for alloys. Some highly relevant quantities for optoelectronic device design and operation are the bandgap, the effective mass, and mobility (or microscopic scattering rates).

Localization and Light Emission

A key feature of light-emitting as well as energy-conversion devices is the carrier recombination dynamics in the presence of dislocation-related or other type of traps. The overall recombination rate is given by the sum of radiative and non-radiative recombination.

High levels of alloying in InGaN can lead to significant departures from the Bloch picture of delocalized states for non-alloyed crystals. There is evidence from numerical calculations in the literature that the hole states can become significantly localized. The radiative recombination rates of localized hole states can vary a lot from those of delocalized Bloch-type holes. Theoretical calculations can become exceedingly difficult in many cases and careful experiments are often needed to understand the effect of carrier localization on radiative recombination, non-radiative recombination, and internal quantum efficiency (IQE) in light emitters.

When high levels of alloying are used in quantum wires and QDs, there can be significant departures from the effective mass description of electron states and hole states in terms of envelope functions. Wire-to-wire variations and dot-to-dot variations in highly alloyed InGaN quantum wires and QDs can manifest in the emission characteristics.

As will be discussed in Chapter 5, the electron and hole distributions in QWs, quantum wires, and QDs in light emitters (lasers and LEDs) are determined by electron quasi-Fermi levels and hole quasi-Fermi levels. When localization of holes occurs we have to think in terms of localized hole states being occupied in accordance with the hole quasi-Fermi level. Built-in electric fields due to polarization lead to localized holes being located away from the electrons, similar to the cases when the alloy disorder and localization effects can be neglected. Therefore, localization does not solve the polarization-induced Stark shifts in InGaN QWs, quantum wires, and QDs. The same is true of AlGaN quantum wells in UV emitters.

Staggered InGaN Quantum Wells

The use of staggered configurations alleviates the band-bending problem inside the quantum wells. The quantum-well composition undergoes a step shift in these designs and a higher indium composition layer is followed by a lower indium composition layer. Since the polarization discontinuity at the lower indium–higher indium layers interface has the opposite sign of that at the GaN/InGaN interface looking downwards from the top, the electric field undergoes a step shift of the opposite sign. This is the origin of the lower band bending in these staggered quantum wells. Figure 4.16 shows staggered well band profiles with lower overall band bending.

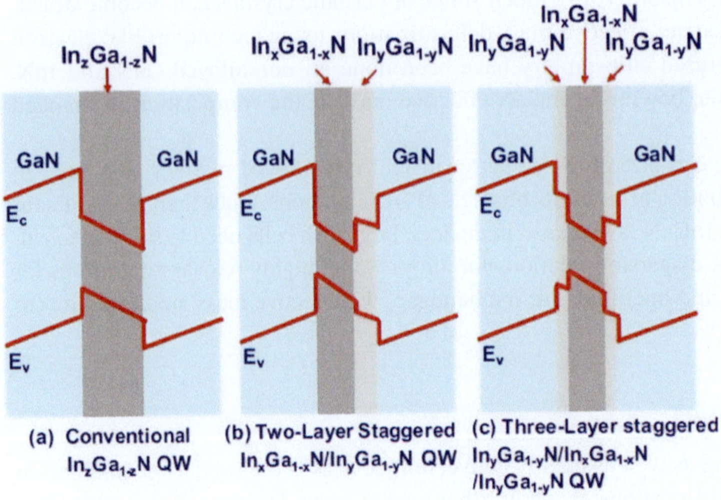

Figure 4.16 Band profiles of non-staggered and staggered QWs (Reference [7]).

$$\sigma_1 = P_{1\ spontaneous} + P_{1\ piezoelectric} - P_{2\ spontaneous} - P_{2\ piezoelectric} \tag{4.18}$$

$$\sigma_2 = P_{1\ spontaneous} + P_{1\ piezoelectric} - P_{2\ spontaneous} - P_{2\ piezoelectric} \tag{4.19}$$

The electric field is given by the continuity of the electric displacement vector at the interface in the absence of free interface sheet charge

$$D_1 = D_2 \tag{4.20}$$

$$\varepsilon_1 E_1 + P_{1\ spontaneous} + P_{1\ piezoelectric} = \varepsilon_2 E_2 + P_{2\ spontaneous} + P_{2\ piezoelectric} \tag{4.21}$$

Similar equations hold for layers two and three as well. The magnitude of the electric field discontinuity at typical *c*-axis GaN/InGaN heterojunctions is of the order of $10^6 \mathrm{V cm^{-1}}$. Since in staggered well designs $\sigma 1$ and $\sigma 2$ have opposite signs, the total band bending is less than in a single quantum-well design of same width. This leads to better overlap between electron and hole wave functions and higher gain spectra, as shown in Figure 4.17. Staggered quantum-well designs can be used in combination with substrate engineering for obtaining even more improved results. Staggered designs are potentially extendable to quantum wire- and QD-based light emitters also.

The interest in staggered well designs stems from the fact that these designs are relevant to *c*-axis growth. Non-polar *a*-plane and *m*-plane growths typically present problems with crystal quality, dislocation density, etc., even though they eliminate polarization along the growth axes. Staggered well designs can provide a good compromise between polarization reduction requirements and dislocation reduction needs.

Triangular Barrier Quantum Wells

Triangular barrier quantum wells refers to the use of composition grading in the barrier layers of the quantum well. For example, the well region can be made of InGaN of a definite In composition, and the barrier layers have grading from GaN down to InGaN with some high In composition and then back again to GaN. This achieves some band bending and electric field at the InGaN/GaN

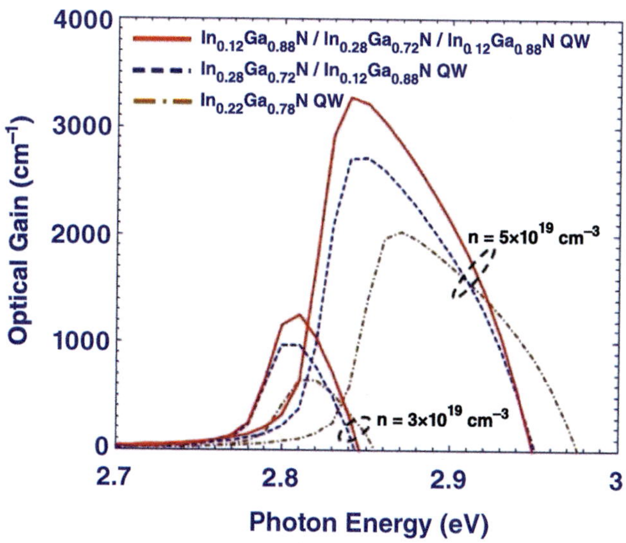

Figure 4.17 Optical gain spectra for non-staggered and staggered quantum wells (Reference [7]).

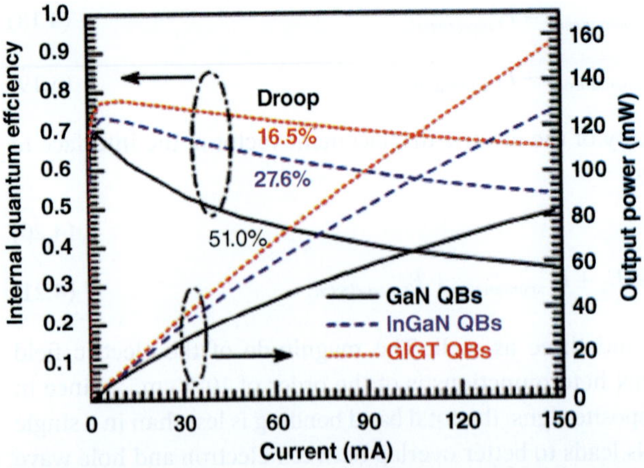

Figure 4.18 LED parameter improvements using a triangular barrier. From Reference [8], used with permission.

quantum well–barrier interface because composition grading leads to polarization grading, and polarization grading is a source of bound charge density. This bound charge density leads to extra electric field through the Poisson equation. The conduction- and valence-band profiles tend to acquire a triangular shape in the barrier, hence the name triangular quantum well. This electric field in the barrier tends to cancel the interface charge at the GaN/InGaN barrier–well interface. The net result is a much smaller built-in electric field in the well layer and a reduction in the QCSE. Figure 4.18 shows the improvement in LED characteristics like IQE (defined in Chapter 5) and output power due to the use of a triangular barrier.

Piezoelectric Charge Fluctuations in InGaN/GaN Quantum Wells

Since the InGaN/GaN quantum well has random alloy InGaN in the well layer, there will be piezoelectric charge fluctuations in the well layer. This is because the actual potential (or pseudopotential) at any given lattice site will be different from the average potential. This leads to local strain fluctuations, and as a result, there will be local fluctuations of piezoelectrically induced charge. The InGaN system has much higher piezoelectric coefficients compared to systems like InGaAs or AlGaAs. Therefore, this local piezoelectric charge fluctuation will be much higher in InGaN quantum wells. This leads to substantial piezoelectric potential fluctuations in these wells.

As the use of the InGaN system is explored to cover wide emission ranges like blue, green, yellow, and red the InGaN composition in well layers must be varied over a wide range. This means that an accurate knowledge of the piezoelectric charge fluctuations becomes necessary to know if these charge fluctuations have significant effect on electron and hole states in the quantum well.

Figure 4.19 shows some strain variations in a typical quantum well and the calculated piezoelectric potentials in the well. These potential fluctuations have been calculated for a cubic InGaN/GaN quantum well. However, similar magnitudes can be expected for wurtzite InGaN/GaN quantum wells also. The piezoelectric fluctuations can lead to carrier localization as well as modification of interband matrix elements and radiative efficiencies. First order calculations and design of quantum-well lasers and LEDs are typically done without these piezoelectric fluctuations. However, as the InGaN composition is varied to get red-yellow-green-blue emissions, these fluctuations can become increasingly important.

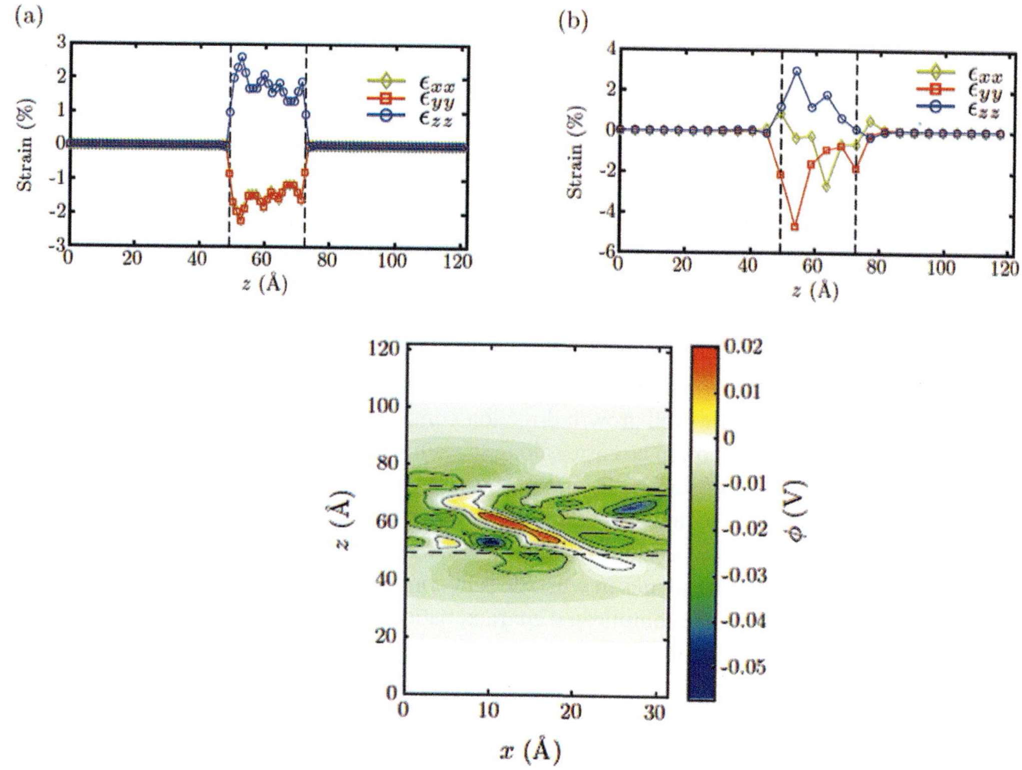

Figure 4.19 Strain fluctuations (top) and piezoelectric potential variation (bottom) in InGaN quantum wells. From Reference [9], used with permission.

Nanowires and Quantum Wires

Nanowires are structures with nanoscale lateral dimensions in the form of rods or cylinders that can be grown by MBE or MOCVD in many circumstances. The definition of a quantum wire is quantum mechanical in nature and has to do with the fact that the thermal deBroglie wavelengths of electrons in both lateral directions become comparable to the lateral dimensions. Nanowires can be quantum wires if this condition is satisfied. Even when the nanowires do not meet the quantum wire definition they still offer many advantages in optoelectronic devices because they give reduced piezoelectric charges and better electron-hole overlaps, etc.

The major advantage of nanowires is that they can offer lower built-in fields (lower spontaneous and piezoelectric polarization-induced charges) even with the technologically easy *c*-axis growth, so that there is better overlap of electron and hole wavefunctions for more efficient emission. Quantum wires, on the other hand, rely on modifying the DOS to give lower threshold currents and higher modulation frequencies. Typically, quantum wires need to be obtained by lithography while nanowires can be grown using the right substrates and growth conditions. Most of the work on group III-Nitride nanowires in the literature is for *c*-axis growth.

Non-polar Plane Quantum Wells, Wires, and Dots

There are planes in wurtzite semiconductors that are perpendicular to the *c*-axis along which there is no spontaneous or piezoelectric polarization. These are called *a*-plane and *m*-plane, respectively.

Growing quantum wells along these non-polar axes leads to quantum wells without built-in electric fields, that is, there is no QCSE in these structures. The electron and hole wavefunctions no longer have large spatial separation in these wells. As will be discussed in Chapter 5, this leads to better IQEs in LEDs and lasers built along these directions.

Non-polar plane quantum wires are expected to show changes similar to non-polar plane quantum wells. Non-polar plane QDs are more difficult to understand because some facets of the QDs will be along or will have components along the *c*-axis. That means these QDs may not have built-in fields along the growth axis, but they will have built-in fields perpendicular to the growth axis.

Semi-polar Plane Quantum Wells, Wires, and Dots

There are other planes that are intermediate between polar and non-polar planes. They have less polarization along the quantum-well growth direction compared to the polar planes. Controlling quality of epitaxial growth can often be easier along the semi-polar planes compared to non-polar planes. Effective masses along semi-polar growth axes can vary significantly from those along polar axes. That means device parameters can vary significantly from polar axes not only due to the reduced built-in electric fields but also due to the changes in other structure parameters. Figure 4.20 shows non-polar and semi-polar planes in wurtzite semiconductors.

Semi-polar plane QDs can have complicated characteristics because they can have polarizations components perpendicular to many faces of the dot. This means that the calculation of dot energy levels for comparison with experiments becomes a potential problem. The reader can find details of non-polar and semi-polar group III-Nitride devices in References [11–34] as well as References [35–38]. The effect of substrates on device performance can be found in References [39–45].

Delta Quantum Wells

Delta wells refer to quantum wells where the well layer is a few atomic layers thick. One of the major reasons for interest in delta wells is the fact that when the well layer is ultra-thin, the separation of electron and hole wavefunctions due to polarization-induced electric fields is reduced. This can potentially lead to better emission efficiencies and to the reduction of red shifts in emission wavelengths. It must be kept in mind that delta wells require precise control of growth rates and crystal quality so that the growth calibration and optimization becomes harder compared to conventional quantum wells.

Because the delta wells are so thin, the use of effective mass approximation and k.p parameters derived for bulk material can lead to large errors. It becomes necessary to do atomistic calculations

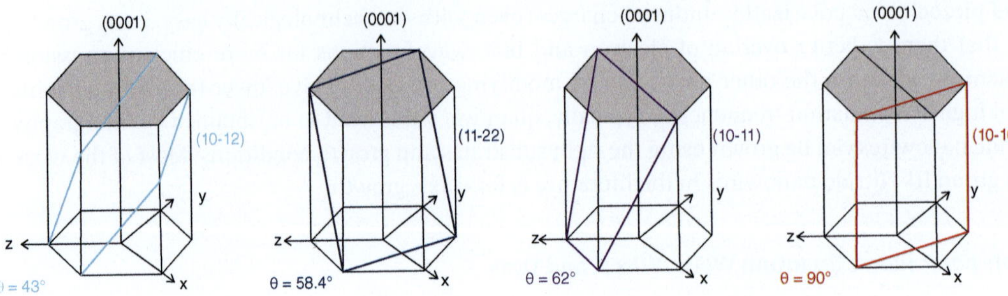

Figure 4.20 Non-polar and semi-polar planes. From Reference [10], used with permission.

where the atomically thin nature of the well layer is taken into account. Another thing that must be considered in delta-well designs is the fact that the electromagnetic modes can look very different from conventional quantum wells due to the ultra-thin nature of the delta well. In fact, parameters like frequency dependent dielectric function may not be well-defined for delta wells and atomistic calculations that consider the well and barrier as a whole system may become necessary. Furthermore, delta quantum wells can have large sub-band energy level magnitudes measured from the top of the valence band or the bottom of the conduction band. This can lead to carrier confinement problems unless very careful optimization is done.

References

1 Chuang, S.L. and Chuang, S. (2009). *Physics of Optoelectronic Devices*. Wiley.

2 Chang, J.R. et al. (2012). Fabrication and luminescent properties of core-shell InGaN/GaN multiple quantum wells on GaN nanopillars. *Appl. Phys. Lett.* 100.

3 Adelmann, C. et al. (2000). Self-assembled InGaN quantum dots grown by molecular-beam epitaxy. *Appl. Phys. Lett.* 76 (12).

4 Soto Rodriguez, P.E.D. et al. (2015). Stranski-Krastanov InN/InGaN quantum dots grown directly on Si(111). *Appl. Phys. Lett.* 106.

5 Winkelnkemper, M. et al. (2006). Interrelation of structural and electronic properties in In*x*Ga1−*x*N/GaN quantum dots using an eight-band k·p model. *Phys. Rev. B* 74.

6 Wei, X. et al. (2022). Near-infrared electroluminescence of AlGaN capped InGaN quantum dots formed by controlled growth on photoelectrochemical etched quantum dot templates. *Photonics Res.* 10 (1).

7 Zhao, H. and Tansu, N. (2010). Optical gain characteristics of staggered InGaN quantum well lasers. *J. Appl. Phys.* 107.

8 Cheng, L. et al. (2018). Suppressed polarization effect and enhanced carrier confinement in InGaN light emitting diodes with GaN/InGaN/GaN triangular barriers. *J. Appl. Phys.* 123.

9 Shireen, T. et al. (2021). Atomistic analysis of piezoelectric potential fluctuations in zinc-blende InGaN/GaN quantum wells: a Stillinger-Weber potential based analysis. *Phys Rev B* 103.

10 Roy, S. et al. (2020). Numerical investigation into optoelectronic performance of InGaN blue laser in polar, non-polar and semipolar crystal orientation. *Crystals*. 10 (11).

11 DMikou, I. et al. (2020). InGaN quantum dots studied by correlative microscopy techniques for enhanced light-emitting diodes. *ACS Appl. Nano Mater.* 3 (10).

12 Xing, K., Tseng, C., Wang, L. et al. (2019). Semi-polar (112¯2112¯2) GaN epitaxial films with significantly reduced defect densities grown on m-plane sapphire using a sequence of two in situ SiNx interlayers. *Appl. Phys. Lett.* 114.

13 Chung, R.B. et al. (2010). Electroluminescence characterization of (20(2) over-bar1)InGaN/GaN light emitting diodes with various wavelengths. *Jpn. J. Appl. Phys.* 49 (7R).

14 Lin, H.-W., Lu, Y.-J., Chen, H.-Y. et al. (2010). InGaN/GaN nanorod array white lightemitting diode. *Appl. Phys. Lett.* 97.

15 Wang, L., Wang, L., Yu, J.D. et al. (2019). Abnormal Stranski-Krastanov mode growth of green InGaN quantum dots: morphology, optical properties, and applications in light-emitting devices. *ACS Appl. Mater. Interfaces* 11 (1).

16 Lee, L.K., Aagesen, L.K., Thornton, K., and Ku, P.-C. (2014). Origin of broad luminescence from sitecontrolled InGaN nanodots fabricated by selective-area epitaxy. *Phys. Status Solidi A* 211 (3).

17 Liu, G., Zhao, H., Zhang, J. et al. (2011). Selective area epitaxy of ultra-high density InGaN quantum dots by diblock copolymer lithography. *Nanoscale Res. Lett.* 6.

18 Philip, M.R., Bui, T.H.Q., Djavid, D.D.C.M. et al. (2018). Molecular beam epitaxial growth and device characterization of AlGaN UV-B nanowire light-emitting diodes. *J. Adv. Opt. Photonics* 1 (1).

19 Philip, M.R., Choudhary, D.D., Djavid, M. et al. (2017). High efficiency green/yellow and red InGaN/AlGaN nanowire light-emitting diodes grown by molecular beam epitaxy. *J. Sci. Adv. Mater. Devices* 2 (2).

20 Nguyen, H.P.T., Cui, K., Zhang, S. et al. (2012). Controlling electron overflow in Phosphor-free InGaN/GaN nanowire white light-emitting diodes. *Nano Lett.* 12 (3).

21 Senichev, A. et al. (2019). Impact of growth conditions and strain on indium incorporation in non-polar m-plane (101¯0101¯0) InGaN grown by plasma-assisted molecular beam epitaxy. *APL Mater.* 7.

22 Reilly, C.E., Bonef, B., Nakamura, S. et al. (2019). Characterization of InGaN quantum dots grown by metalorganic chemical vapor deposition. *Semiconductor Sci. Technol.* 34.

23 Young, E.C. et al. (2010). Critical thickness for onset of plastic relaxation in (11–22) and (20–21) semipolar AlGaN heterostructures. *Appl.Phys. Express* 3.

24 Rass, J., Wernicke, T., Scheibenzuber, W.G. et al. (2010). Polarization of eigenmodes in laser diode waveguides on semipolar and nonpolar GaN. *Phys. Status Solidi Rapid Res. Lett.* 4.

25 Khoury, M. et al. (2020). Polarized monolithic white semipolar (20-21) InGaN light-emitting diodes grown on high quality (20-21) GaN/sapphire templates and its application to visible light communication. *Nano Energ.* 67.

26 Nguyen, H.P.T., Cui, K., Zhang, S. et al. (2011). Full-color InGaN/GaN dot-in-a-wire light emitting diodes on silicon. *Nanotechnol.* 22 (44).

27 Wang, Q., Nguyen, H.P.T., Cui, K., and Mi, Z. (2012). High efficiency ultraviolet emission from AlxGa1 _ xN core-shell nanowire heterostructures grown on Si(111) by molecular beam epitaxy. *Appl. Phys. Lett.* 101.

28 Lee, L.K. and Ku, P.-C. (2012). Fabrication of site-controlled InGaN quantum dots using reactive-ion etching. *Phys. Status Solidi C* 9.

29 Liu, J., Jia, Z., Ma, S. et al. (2018). Enhancement of carrier localization effect and internal quantum efficiency through In-rich InGaN quantum dots. *Superlattices Microstruct.* 113.

30 Zhang, H., Li, H., Li, P. et al. (2020). Room-Temperature continuous-wave electrically driven semipolar (2021¯¯¯¯2021¯) Blue laser diodes heteroepitaxially grown on a Sapphire substrate. *ACS Photonics* 7 (7).

31 Tourbot, G. et al. (2012). Growth mechanism and properties of InGaN insertions in GaN nanowires. *Nanotechnol.* 23.

32 Shen, S. et al. (2020). Semi-polar InGaN-based green light-emitting diodes grown on silicon. *Phys. Status Solidi A* 217 (7).

33 Koslow, I.L. et al. (2014). Onset of plastic relaxation in semipolar (11–22) InxGa1-xN/GaN heterostructures. *J. Cryst. Growth* 388.

34 Zhang, Y. et al. (2016). Microstructure investigation of semi-polar (11–22) GaN overgrown on differently designed micro-rod array templates. *Appl. Phys. Lett.* 109.

35 Bai, J. et al. (2015). (11–22) Semipolar InGaN emitters from green to amber on overgrown GaN on micro-rod templates. *Appl. Phys. Lett.* 107.

36 Wang, T. (2016). Topical review: development of overgrown semi-polar GaN for high efficiency green/yellow emission. *Semicond. Sci. Technol.* 31 (9).

37 Romanov, A.E. et al. (2011). Basal plane misfit dislocations and stress relaxation in III-nitride semipolar heteroepitaxy. *J. Appl. Phys.* 109.

38 Yu, X., Hou, Y., Shen, S. et al. (2016). Semi-polar (11–22) GaN grown on patterned (113) Si substrate. *Phys. Status Solidi C* 13.

39 Wang, L. et al. (2018). Abnormal Stranski–Krastanov mode growth of green InGaN quantum dots: morphology, optical properties, and applications in light-emitting devices. *ACS Appl. Mater. Interfaces* 11 (1).

40 Waag, A., Wang, X., Fundling, S. et al. (2011). The nanorod approach: gaN NanoLEDs for solid state lighting. *Phys. Status Solidi C* 8.

41 Cui, K., Fathololoumi, S., Golam Kibria, M. et al. (2012). Molecular beam epitaxial growth and characterization of InN/InxGa1 _ xN core/shell nanowire heterostructures on Si(111) substrates. *Nanotechnol.* 23 (8).

42 Meissner, C., Ploch, S., Pristovsek, M., and Kneissl, M. (2009). Volmer-Weber growth mode of InN quantum dots on GaN by MOVPE. *Phys. Status Solidi* 6 (52).

43 Sekiguchi, I., Kishino, K., and Kikuchi, A. (2010). Emission color control from blue to red with nanocolumn diameter of InGaN/GaN nanocolumn arrays grown on same substrate. *Appl. Phys. Lett.* 96.

44 Nguyen, H.P.T., Zhang, S., Cui, K. et al. (2012). High efficiency InGaN/GaN dot-in-a-wire red light emitting diodes. *IEEE Photon Tech. Lett.* 24 (4).

45 Sawicka, M., Feduniewicz-Żmuda, A., Kryśko, M. et al. (2017). Indium incorporation in semipolar (202¯1202¯1) and nonpolar (101¯0101¯0) InGaN grown by plasma assisted molecular beam epitaxy. *J. Cryst. Growth* 459.

5

Light-emitting Diodes and Lasers

Blue, Green, and Ultraviolet (UV) LEDs

Blue LEDs

Since the blue end of the spectrum can be reached using low In-content indium gallium nitride (InGaN) and other semiconductor material systems usually give emission up to green (for example, zinc selenide [ZnSe]-based light sources), group III-Nitride blue light-emitting diodes (LEDs) are of great technological interest.

Early designs of multi quantum-well LEDs using *c*-axis gallium nitride (GaN)/InGaN quantum wells faced the problem of serious red shift due to polarization-induced interface charges and built-in electric fields. Carrier screening cannot fully alleviate this problem since electrons and holes tend to be present in almost similar concentrations when the LED is biased to give a high optical output. Non-polar plane LED designs or specialized designs like staggered-well designs, both of which are covered later in the chapter (in the context of LEDs, lasers, or both), become necessary to overcome the red shift as well as to improve overlap between electron and hole wave functions (needed to improve quantum mechanical transition amplitudes between conduction-band and valence-band states) to improve quantum efficiencies and power efficiencies.

Green LEDs

The InGaN composition can be varied to get emission in the green part of the optical spectrum as well (in fact, in principle, the emission can be varied from the ultraviolet [UV] through the optical to infrared). While ZnSe already provides green light-emitting devices, the ability to get red, green, and blue emission from the same material system can be a great advantage for some applications like optoelectronic integrated circuits and quantum computing using photon states.

The ability to get high-efficiency green light diodes using InGaN quantum wells depends, like in the blue case, on the ability to change the composition of the barrier layer to obtain small magnitudes of polarization-induced bound interface charges. The major challenge with getting green LEDs in the InGaN material system is built-in fields due to polarization-induced charges. Green LEDs require a higher In concentration in the InGaN quantum well compared to blue LEDs. This leads to larger lattice mismatch with the GaN barrier layers and much higher piezoelectric polarization, which leads to interface charges that cause high built-in fields and Stark shifts in the emission as well as reduced overlap between electrons and holes.

Group III-Nitride Semiconductor Optoelectronics, First Edition. C. Jayant Praharaj.

The larger lattice mismatch for green LEDs also means that there is a larger chance of getting misfit dislocations. Although the InGaN well-layer thickness can be kept under critical thickness to prevent dislocation formation, the exact optimization between getting the right wavelength and reducing dislocations can be challenging.

UV LEDs

UV LEDs require GaN or even aluminum gallium nitride (AlGaN) light-emitting layers. This makes the ability to get good ohmic contacts to the *p* and *n* surrounding layers very difficult since both Schottky barrier heights and effective masses are large. Thin polarization-engineered caps can be used to lower the tunneling barrier width for the same Schottky barrier heights and get good non-annealed ohmic contacts to *p*- and *n*-GaN or AlGaN. While AlGaN LEDs cannot cover all ranges of the UV spectrum (for example, they cannot emit in some wavelengths that free electron lasers can emit in), they can still be useful for a broad array of applications.

Polarization-induced interface charges at quantum-well interfaces or multi quantum-well interfaces also provide an engineering challenge for obtaining high-efficiency UV LEDs, similar to green LEDs, because the quantum-confined Stark effect due to the built-in electric field leads to a severe red shift. That means AlGaN with reasonably high Al content can become necessary as the quantum-well layer to increase the bandgap and compensate for the red shift, with AlGaN of even higher Al content needed as the barrier layers. This creates technological difficulties because of several reasons. *p*-Type doping of AlGaN is very difficult when the Al percentage is too high because the hole masses become very high, and this leads to very large ionization energies for acceptors like Mg. In GaN the acceptor ionization energy can become as high as 200–220 meV. In AlGaN with a high Al percentage this number can become as high as 300–320 meV. This leads to a low ionization percentage for the Mg acceptors. The ionization ratio can be as low as 0.005, which means that even when the physical doping density is as high as 10^{20} per cm^3 the hole density in the *p*-type AlGaN is only 5×10^{17} per cm^3. This leads to high series resistances. Further, when AlGaN with a high Al percentage is used as the *p*-type barrier layer it is difficult to get ohmic contacts because the hole effective masses in the AlGaN layer become as high as 1.5–1.7, and this suppresses hole tunneling, which is needed to obtain low contact resistances at *p*-type ohmic contacts. That means that lower bandgap transition layers are needed to obtain the ohmic contact. However, there tends to be a fairly large barrier for hole flow from the transition layer to the barrier layer in this case. This barrier can become as high as 1 eV or similar values. That means the series resistance plus the ohmic contact resistance remains high.

One way to overcome the Stark effect without going to very high Al percentage AlGaN active layers is to use an ultra-thin quantum-well layer as the active layer of the LED. An ultra-thin quantum-well layer reduces the Stark shift despite the large polarization-induced interface charges. This also leads to better overlap between electron and hole wave functions. However, LEDs with ultra-thin quantum-well layers (also called delta wells), have the problem of beam spreading due to diffraction because the active layer thickness has to be much lower than the free-space wavelengths of the emitted light. Therefore, careful optimization is needed to obtain a good compromise between power efficiencies and optical emission profiles.

AlGaN UV LEDs grown on sapphire, silicon carbide (SiC), or Si substrates have the usual problems with dislocations arising from lattice mismatches and the propagation of dislocations into the device active layers. Aluminum nitride (AlN) substrates are potentially attractive for reducing lattice mismatch with substrate in AlGaN UV LEDs. This leads to lower dislocation densities, lower non-radiative recombination rates, and better quantum efficiencies. Homoepitaxial AlN barrier layers and pseudomorphic AlGaN barrier layers on AlN substrates are choices for obtaining high quality UV LEDs.

Ultra-thin GaN/AlN Heterostructure for UV LEDs

AlGaN UV emitters have a severe problem of lattice mismatch with typical substrates. The use of ultra-thin GaN/AlN heterostructures is a potential solution to this problem. For example, some UV LED designs use a delta-GaN quantum well or delta-GaN/AlN superlattices. Band structure calculations of ultra-thin GaN/AlN superlattices shows that the bandgap remains direct in the superlattice. This means that efficient light emission is possible while also solving lattice-mismatch problems because GaN has better lattice matching with many substrates compared to AlN. 2-ML GaN/AlN heterostructures have a bandgap of 4.65 eV and 1-ML GaN/AlN heterostructure have a bandgap of 5.45 eV.

Along with the use of ultra-thin GaN/AlN layers to obtain efficient emission and high internal quantum efficiency for UV LEDs, polarization-induced doping can be used to obtain high doping concentrations in AlGaN layers. Since composition grading leads to polarization grading and polarization grading leads to bound charge densities that must be compensated by mobile carrier charges to maintain overall charge neutrality in the device, we can use composition grading to obtain layers where the hole densities and electron densities, respectively, are high despite high ionization energies for acceptors and donors. The other variation from continuous grading is the use of superlattice layers. Polarization-induced doping is robust against temperature variations over a wide range of temperature. This can enable cryogenic temperature operation of UV LEDs.

Ultra-thin GaN/AlN heterostructures also have better overlap between electron wave functions and hole wave functions compared to typical group III-Nitride structures. This is a major component of improved internal quantum efficiencies.

By making high-precision changes in the GaN layer (monolayer precision), the UV emission wavelength can be easily tuned over a wide range. For example, the emission range can be tuned from 232 nm to 270 nm by making GaN thickness variations.

Indium Gallium Nitride UV LEDs

UV emission in the range of 365–420 nm can be obtained using optimized InGaN/GaN LEDs. However, potential barriers for carriers are typically lower compared to the GaN/AlN UV LEDs. This leads to carrier-confinement problems and overflow processes and necessitates careful polarization engineering involving layer compositions and widths.

The carrier-confinement problems necessitate electron-blocking layers. A typical electron-blocking layer (EBL) consists of *p*-type AlGaN/GaN superlattices. The superlattice leads to improved acceptor activation of the *p*-type doping and the AlGaN leads to a higher bandgap necessary to block electrons. EBLs are discussed in greater detail later in the chapter. Figure 5.1 shows typical group III-Nitride UV LED characteristics.

Blue-Green-Yellow-Red MicroLEDs

There is a lot of interest in making blue-green-yellow-red microLEDs for use in display technologies. There are obvious advantages of making these LEDs from the same material system. So, the

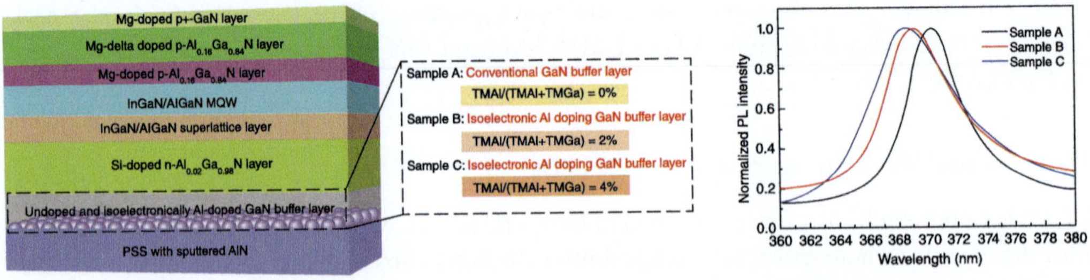

Figure 5.1 InGaN UV LED characteristics, after Reference [1], used with permission.

InGaN system has been explored to achieve production of these microLEDs. The use of quantum dots (QDs) to reduce built-in electric fields due to reduced polarization-induced charges has become the most promising way to achieve these goals. Stranski–Krastanow mode QDs using either molecular beam epitaxy (MBE) or metal-organic chemical vapor deposition (MOCVD) have been used.

Another method is the use of nanowire LEDs to obtain red-emitting microLEDs. The other material system used to obtain red-emitting microLEDs is aluminum gallium indium phosphide (AlInGaP). However, there is a sharp efficiency cliff below a 100 μm lateral size of LEDs, as shown in Figure 5.2. Below this cliff, the external quantum efficiency (defined later in the chapter) drops drastically.

Similar behavior is also observed for conventional InGaN/GaN LEDs in the red and yellow wavelength range. However, nanowire InGaN microLEDs can give emission in the red with better efficiency, although more research is needed to improve the efficiencies to acceptable levels. Compared to AlInGaP, GaInN based heterostructures offer stronger carrier confinement, smaller carrier diffusion lengths, and a lower level of surface recombination velocity. The lattice mismatch between indium nitride (InN; 3.54 Å) and GaN (3.18 Å) is quite large, so that there is a large lattice mismatch between InGaN and GaN when large In concentrations have to be used, as is required for red and yellow LEDs. This leads to large dislocation densities in conventional InGaN/GaN LEDs for red and yellow wavelengths as well as high rates of non-radiative recombination, which decreases the radiative efficiency or internal quantum efficiency (defined in the next section). Nanowire-based InGaN LEDs provide much lower dislocation densities and therefore help overcome the efficiency cliff.

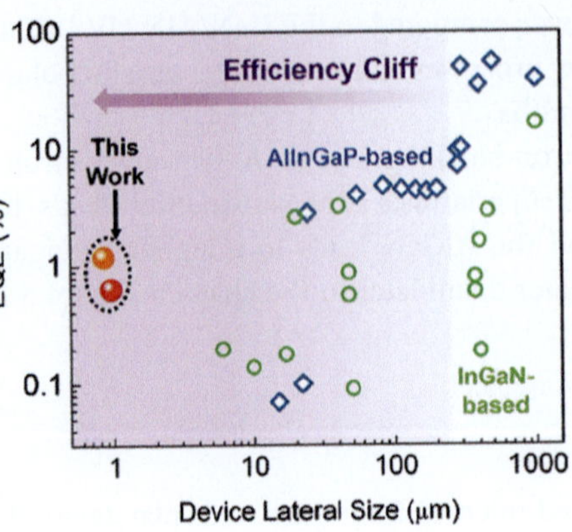

Figure 5.2 Efficiency cliff in micro LEDs and comparison of nitrides with phosphides. After Reference [9], used with permission.

Light-emitting Diode Basic Operating Principles

Recombination Mechanisms and Quantum Efficiency

The simplest LED structure consists of a *p-n* junction where the radiative recombination of injected electrons and holes leads to light emission. More advanced LEDs use heterojunctions like quantum wells to improve the emission characteristics. It must be kept in mind that non-radiative recombination processes occur along with radiative recombination. The radiative recombination in LEDs is a spontaneous emission process because the optical feedback is not sufficient to build up the optical intensity in the main device layer to provide gain via stimulated emission (present in lasers due to strong feedback and enough current density to overcome mirror losses, which leads to positive gain and build up of optical intensity). Some LEDs can operate as superluminescent LEDs, which are intermediates between normal LEDs and lasers. In superluminescent LEDs there is some stimulated emission, but the optical gain is not enough to overcome facet (mirror) losses so that laser operation is not achieved.

The total recombination rate in the LED active layer is the sum of the radiative and non-radiative recombination rates

$$\frac{1}{\tau_{rec}} = \frac{1}{\tau_r} + \frac{1}{\tau_{nr}} \tag{5.1}$$

where τ_r and τ_{nr} are the radiative and non-radiative recombination rates, respectively. The non-radiative recombination rate can be due to deep traps in the semiconductor, which result in Shockley–Read–Hall (SRH)-type recombination, or it can be due to Auger recombination in which an electron recombines with a hole and gives its energy to another electron.

A useful Figure of merit for an LED is the internal quantum efficiency, which is defined as

$$\eta_i = \frac{Number\ of\ photons\ emitted\ internally\ per\ unit\ time}{Number\ of\ carriers\ injected\ across\ the\ junction\ per\ unit\ time} \tag{5.2}$$

Since the injected electron current contributes to the sum of radiative and non-radiative recombination rates, it can be shown that

$$\eta_i = \frac{R_r}{R_r + R_{nr}} = \frac{\tau_{nr}}{\tau_r + \tau_{nr}} \tag{5.3}$$

The external quantum efficiency of the LED depends on the internal quantum efficiency and other structure-related factors like reflectivity of the device facets, etc.

The radiative recombination rate is of the form

$$R_r = Bnp \tag{5.4}$$

where n and p are the electron and hole concentrations, respectively. The value of B is typically $10^{-10} cm^3 sec^{-1}$ for direct bandgap materials. Figure 5.3 shows typical recombination mechanisms relevant to the quantum efficiency of devices.

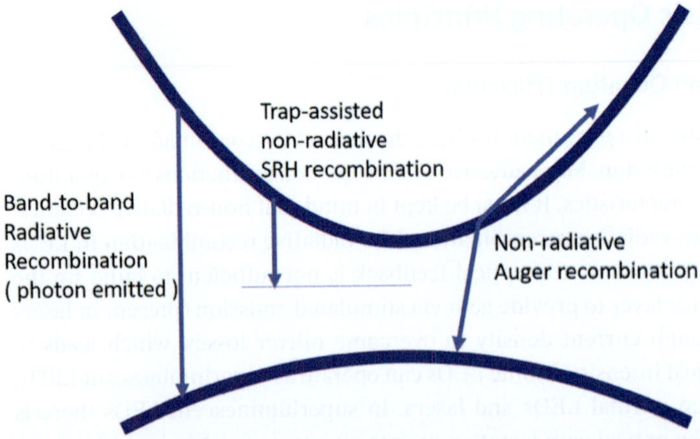

Figure 5.3 Different kinds of recombination processes in group III-Nitride devices.

Frequency Response of Light-emitting Diode

The rate at which the LED can be turned on or off is often a useful parameter because of potential use in data transmission. Solving the device current continuity equations gives the cut-off frequency for the LED as

$$f_t = \frac{1}{2\pi\tau_{rec}} \tag{5.5}$$

Blue, Green, and UV Lasers

Blue, Green, and Green Laser Device Challenges

Blue, green, and UV lasers made from group III-Nitride heterojunctions like multi-quantum wells face a lot of similar challenges to LEDs, like severe band bending due to polarization-induced interface charges and built-in electric fields. Therefore, a lot of the solutions like staggered multi-quantum wells and polarization engineering using ternary and quaternary alloy compositions tend to be similar to those used to improve LEDs. The strain in the quantum well and the barrier layers become crucial to the design because the piezoelectric polarization can be of the same order as the spontaneous polarization in non-optimized designs.

Sub-bands in Quantum Wells

The effective mass approximation is used to obtain the quantized energy levels perpendicular to the quantum-well layers.

$$-\frac{\hbar^2}{2\,m_\perp^*}\nabla^2\Psi + V\Psi = E\Psi \tag{5.6}$$

where ħ is the reduced Planck's constant (Planck's constant divided by 2Π), $m_\perp^*$ is the electron effective mass perpendicular to the quantum well, and V is the slowly varying potential above the crystal Hamiltonian. The eigenvalues of this equation are the allowed energy levels for perpendicular quantization. In the transverse direction, the usual E-$\boldsymbol{k_x}$,$\boldsymbol{k_y}$ dispersion relation holds. The eigenfunctions that correspond to these eigenvalues are the envelope functions that need to be multiplied with the cell-periodic part to give the total wave function.

$$\psi_c(\boldsymbol{r}) = u_c(\boldsymbol{r})\frac{exp\left(i\boldsymbol{k}_t.\boldsymbol{r}_t\right)}{\sqrt{A}}f(z) \tag{5.7}$$

The valence band is more complicated due to the presence of three kinds of holes: i) the heavy holes, ii) the light holes, and iii) the spin-orbit split-off holes. Focusing on one type of holes, for example the heavy holes, we have for the valence-band wave function in the quantum well

$$\psi_v(\boldsymbol{r}) = u_v(\boldsymbol{r})\frac{exp\left(ik_t'.\boldsymbol{r}_t\right)}{\sqrt{A}}g(z) \tag{5.8}$$

The exact valence-band wave functions and the valence-band dispersions can be obtained by solving the six-band k.p equation (based on the Luttinger–Kohn Hamiltonian) or by solving the eight-band k.p equation. The larger the bandgap, the more accurate the six-band k.p equation is. When the band bending due to polarization-induced electric fields is too large, valence-band states and conduction-band states get closer in energy and the eight-band method may become necessary. It must be kept in mind that since lasers and LEDs are designed to emit in the design wavelength (and therefore, energy), which typically is above 2 eV (actually well above 2 eV for nitride lasers), the difference between conduction-band and valence-band energies is kept above 2 eV by design. Therefore, the six-band k.p method can give accurate results.

The Population Inversion Condition and the Gain Coefficient

The gain coefficient is the negative of the loss (absorption) coefficient, which is given by

$$\alpha(\hbar\omega) = \frac{\pi e^2}{n_r c\varepsilon_0 m_0^2 \omega}\frac{2}{V}\sum\nolimits_k \left|\boldsymbol{e}.\boldsymbol{p}_{cv}\right|^2 \delta(E_c - E_v - \hbar\omega)(f_v - f_c) \tag{5.9}$$

where q is the magnitude of the electron charge, m_0 is the electron mass in free space, and c and ε_0 are the velocity of light and permittivity in free space, respectively. e is the polarization vector of the optical electric field; n_r is the refractive index. In the momentum matrix element, the cell-periodic part momentum matrix element is multiplied with the overlap integral between conduction and valence-band states to a high level of accuracy. The gain coefficent $g(\hbar\omega)$, being just the negative of $\alpha(\hbar\omega)$, is positive when the Fermi–Dirac occupation factor is higher in the conduction band compared to the valence band for states at the same $\boldsymbol{k}$-vector separated by $\hbar\omega$.

In quantum wells and quantum wires, the matrix element becomes the product of the momentum matrix element between the cell-periodic parts of the Bloch wave functions and the overlap integral between the envelope functions of the conduction-band state and the valence-band state. Large built-in fields reduce the overlap integral by shifting the centers of the conduction-band and valence-band wave functions and therefore reduce the overlap integral. In non-optimized designs, the overlap integral can become as small as 0.01 and can reduce the gain coefficient by a lot.

Bias Conditions and the Gain Coefficient

The gain coefficient also involves the difference between conduction-band Fermi–Dirac occupation factor and the valence-band Fermi–Dirac occupation factor. To get a positive gain of large magnitude, one needs a situation where f_c for states close to the conduction-band minimum becomes much larger than f_v for states close to the valence-band maximum. This is possible only in non-equilibrium and is typically achieved by large positive bias across quantum well (or quantum wire or dot) or double heterojunction laser diodes. When a large positive bias is applied across a quantum well, for example, the quantum-well layer reaches a situation where typical f_c values become much larger than typical f_v values. This happens because the electron quasi-Fermi level and the hole quasi-Fermi level tend to remain flat across the quantum well from n-layer and the p-layer, respectively. This behavior of quasi-Fermi levels follows from solving the device current continuity equations accurately. This configuration of quasi-Fermi levels in the quantum well means that the well attains a large conduction-band electron density and a large valence-band hole density. The gain equation then makes the gain coefficient become large in this configuration. Similar arguments hold for quantum-wire lasers and LEDs as well as for QD lasers and LEDs. Figure 5.4 shows bias conditions and quasi-Fermi levels for lasers and LEDs.

Threshold Current Density and Output Lasing Power

$$\frac{\partial n}{\partial t} = D\nabla^2 n + \frac{J}{\text{qd}} - R(n) \tag{5.10}$$

$$R(n) = A_{nr}n + Bn^2 + Cn^3 + R_{st}N_{ph} \tag{5.11}$$

The $A_{nr}n$ term accounts for non-radiative recombination processes like the trap-assisted SRH recombination processes, the Bn^2 term is due to radiative spontaneous recombination due to direct band-to-band electron-hole recombination, the Cn^3 term is due to Auger recombination involving three-particle carrier scattering processes, and the $R_{st}N_{ph}$ term is due to stimulated emission,

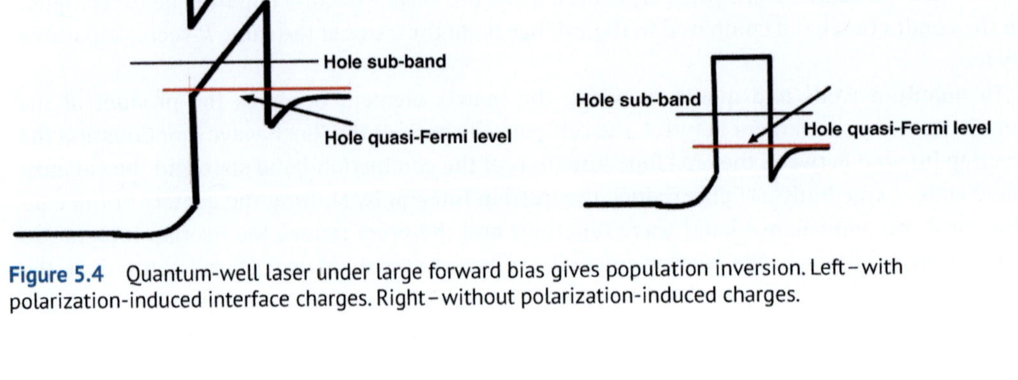

Figure 5.4 Quantum-well laser under large forward bias gives population inversion. Left – with polarization-induced interface charges. Right – without polarization-induced charges.

which is proportional to the photon density on the laser cavity. R_{st} is proportional to the gain coefficient calculated earlier and is also dependent on the level of population inversion in the laser cavity.

$$R_{st} = v_g g(n) \tag{5.12}$$

Below threshold, the steady-state current is given by the equation

$$\frac{J}{qd} = A_{nr} n + Bn^2 + Cn^3 \tag{5.13}$$

$$g(n) = a(n - n_{tr}) \tag{5.14}$$

where n_{tr} is the carrier concentration at which transparency is achieved, that is, the intrinsic gain exactly compensates for loss in the material.

$$g_{th}\Gamma = \alpha_i + \alpha_m \tag{5.15}$$

$$\alpha_i = \alpha_o(1 - \Gamma) + \alpha_g \Gamma \tag{5.16}$$

$$\alpha_m = \frac{1}{2L} \ln\left(\frac{1}{R_1 R_2}\right) \tag{5.17}$$

where R_1 and R_2 are the facet reflectivities.

$$n_{th} = n_{tr} + \frac{\alpha_i + \alpha_m}{\Gamma a} \tag{5.18}$$

$$J_{th} = \frac{qdn_{th}}{\tau(n_{th})} \tag{5.19}$$

where

$$\tau(n_{th}) = A_{nr} + Bn_{th} + Cn_{th}^2 \tag{5.20}$$

As the current is increased so that the electron and hole densities in the active layer increase beyond the threshold densities, stimulated emission process begins. The gain spectrum as a function of n is modified when the photon density in the cavity builds up and gain saturation occurs. The carrier density becomes pinned at the threshold concentration

$$J = qd(A_{nr} n_{th} + Bn_{th}^2 + Cn_{th}^3) + qdR_{st} N_{ph} \tag{5.21}$$

$$J = J_{th} + qdv_g g_{th} N_{ph} \tag{5.22}$$

$$N_{ph} = \frac{1}{qdv_g g_{th}} (J - J_{th}) \tag{5.23}$$

$$\eta_i = \frac{Bn^2 + R_{st}N_{ph}}{A_{nr}n + Bn^2 + Cn^3 + R_{st}N_{ph}} \tag{5.24}$$

The external differential quantum efficiency is given by

$$\eta_e = \eta_i \frac{\ln\left(\frac{1}{R}\right)}{\alpha_i L + \ln\left(\frac{1}{R}\right)} \tag{5.25}$$

Rate Equations and Linear Gain

$$\frac{dN}{dt} = \frac{J}{qd} - \frac{N}{\tau} - vg(N)S \tag{5.26}$$

$$\frac{dS}{dt} = \Gamma vg(N)S - \frac{S}{\tau_p} + \beta R_{sp} \tag{5.27}$$

where $g(N)$ is the gain coefficient in cm^{-1} and β is the spontaneous emission factor.

$$g(N) = g_0 + g'(N - N_0) \tag{5.28}$$

where $g_0 = g(N_0)$ and g' is the differential gain at $N = N_0$.

$$J(t) = J_0 + j(t) \tag{5.29}$$

$$N(t) = N_0 + n(t) \tag{5.30}$$

$$S(t) = S_0 + s(t) \tag{5.31}$$

where the steady-state photon density and carrier concentration are given by

$$S_0 = \frac{\beta R_{sp}}{\frac{1}{\tau_p} - \Gamma v g_0} \tag{5.32}$$

$$N_0 = \tau\left(\frac{J_0}{qd} - v g_0 S_0\right) \tag{5.33}$$

The small signal equations for the photon density and the carrier concentration are

$$\frac{dn(t)}{dt} = \frac{j(t)}{qd} - \frac{n(t)}{\tau} - v(g'S_0 n(t) + g_0 S(t)) \tag{5.34}$$

$$\frac{ds(t)}{dt} = \Gamma v(g'S_0 n(t)) + g_0 s(t)) - \frac{s(t)}{\tau_p} \tag{5.35}$$

Using phasor notation, we obtain

$$n(\omega) = \frac{1}{\Gamma v g' S_0}\left(-i\omega - \Gamma v g_0 + \frac{1}{\tau_p}\right)s(\omega) \tag{5.36}$$

For the frequency response, we obtain

$$\left|\frac{s(\omega)}{j(\omega)}\right| = \frac{\Gamma v g' S_0 / qd}{|D(\omega)|} \tag{5.37}$$

where the denominator is given by

$$D(\omega) = -\omega^2 - i\omega\left(\frac{1}{\tau} + v g' S_0\right) + \frac{v g' S_0}{\tau_p} \tag{5.38}$$

The modulation bandwidth is given by

$$\omega_{3dB} = \sqrt{\frac{v g' S_0}{\tau_p}}$$

The maximum modulation frequency of the laser increases with the differential gain coefficient g'. Quantum wires have higher g' compared to quantum wells. This is due to the lower density of states in the quantum wire compared to quantum wells. The lower density of states means that as n is increased in the conduction band (and p in the valence band), the quasi-Fermi level of electrons moves higher in the conduction band compared to quantum wells. Similarly, the quasi-Fermi level of holes moves lower in the valence band. It must be kept in mind that other factors like optical power density S_0 and photon lifetime τ_p also determine the modulation frequency, which can complicate the speed optimization for any given design.

Similar arguments would lead one to expect naively that QD lasers should have higher modulation frequency compared to quantum-wire lasers. However, there is an added complication in QD lasers due to a phenomenon called phonon blockade. The phonon blockade refers to the fact that electrons in any sub-band level of the QD (and likewise for holes) cannot relax in energy easily due the fact that relaxation by phonon emission cannot happen easily since energy conservation and crystal momentum conservation cannot be satisfied simultaneously. This has to do with the discrete energy spectrum of the QD where energy differences between discrete energy levels may not match phonon frequencies. It must be kept in mind that phonon blockade is to be expected from this kind of basic argument. However, detailed atomistic calculations of QD energy levels and QD phonon spectra can give a better idea of whether phonon blockade happens in any particular case or not.

Characterization of Quantum Wells

Photoluminescence is one of most commonly used methods to understand the carrier quantum states and their correlation with optical transitions. While optoelectronic devices can have extra complications like free-carrier screening, these studies of quantum-well layers provide good insight into the conduction- and valence-band states and the kind of emission that can be expected in devices.

Blue, Green, and Device Laser Materials – Device Considerations

Blue Lasers

The technological considerations for blue lasers tend to be similar to that for blue LEDs. Polarization-induced charges need to be minimized to prevent Stark shifts. Achieving good *p*-type conductivity in the *p*-type barrier (cladding) layer can be a big challenge due to larger hole masses compared to other material systems. Also, getting good *p*-type contacts is difficult due to larger hole masses in this material system, which leads to lower tunneling probabilities at ohmic contacts.

The use of *a*-plane or *m*-plane non-polar growth can eliminate the polarization effects. However, growth on these planes is technologically more difficult and getting low dislocation densities is a challenge. With improvements in growth technologies, these growth planes may become preferred for blue LEDs.

Green Lasers

The InGaN material composition required for green lasing is a high In content, and this leads to much larger polarization-induced charges. Non-polar plane growth and staggered-well designs become much more necessary if one has to solve the Stark shift and the problem of reduced electron-hole overlap. The use of EBLs also become necessary to provide good population inversion in the quantum well.

QD lasers can reduce the built-in fields because there is evidence that QDs have lower interface charge effects. However, getting QDs of the right dimension and In composition is a technological challenge and needs careful optimization.

Green Quantum Dot Lasers

Green QD lasers have been demonstrated with Stranski–Krastanov QD active layers. Several InGaN/GaN QD layers form the active region that mirrors the design on multiple quantum-well (MQW) lasers. QDs can have deeper levels, which prevents electron escape from the active layer.

UV Lasers

The UV laser using quantum well also suffers from polarization-induced Stark shifts. The non-polar plane growth solution can be used with added technological difficulties. Delta quantum-well solutions can be explored where an ultra-thin quantum well is used for a better electron-hole overlap despite polarization-induced charges and band bending. Also, the delta well drastically reduces the Stark shift. However, growing ultra-thin delta wells is a challenge in itself and the growth method can get restricted to MBE only. Also, delta-well designs will lead to much more beam spreading due to diffraction.

The problem of getting *p*-type conductivity in the *p*-type barrier layer and getting good *p*-type ohmic contacts becomes much more difficult because AlGaN barriers become necessary. Doping densities approaching solid solubilities must be used to attain Mott metal-insulator transitions of the *p*-type dopants. If the Mott transition can be achieved, the *p*-type layer will have higher conductivity because the impurity energy level spreads into a band when this happens. The formation of impurity bands brings the impurity levels closer to the valence band. This leads to much higher hole concentration in the valence band and higher *p*-type conductivity. However, there is evidence that trying to push the physical doping concentration in the *p*-type barrier too high leads to vacancy defects that can degrade mobility.

The problem of getting good *p*-type contacts remains because the hole masses are still high, even after the Mott transition. The use of InGaN cap layers (which uses the right sign of polarization-induced charges to reduce the tunneling width for the *p*-contact) becomes necessary in several cases. High-temperature anneals and specialized contact metals is another option, but it provides process integration challenges because other layers can have problems if very-high temperature anneals are used.

Red Quantum Dot Lasers

QD lasers in the red wavelength with high In concentration in the dot layer have been demonstrated. The dot geometry leads to a reduction in polarization-induced charges and built-in fields. This leads to reasonable threshold currents, power efficiencies, and quantum efficiencies.

Substrate Effects with Spontaneous and Piezoelectric Polarization

Wurtzite quantum wells have built-in polarization discontinuities of the spontaneous and piezoelectric variety. Both lasers and LEDs are affected by polarization-induced charges. While non-polar devices with *a*-plane or *m*-plane growth can be built, the substrates tend to be expensive and technological control over dislocations can be problem. Therefore, there is still a lot of interest in building LEDs and lasers along the *c*-axis. Substrate engineering offers a way for alleviating spontaneous and piezoelectric polarization effects for these devices.

InGaN substrates with composition matching or close to the well layer is a way to achieve high emission efficiencies and low threshold currents. The piezoelectric component can be almost completely eliminated using this technique.

In the wurtzite material system, *m*-plane and *a*-plane growth can lead to polarization-free quantum wells and better overlap of conduction- and valence-band wave functions. The modification of dislocation densities can lead to differences in device performance in device structures. As discussed in Chapter 7, *m*-plane growth can happen on *r*-plane sapphire substrate.

Staggered-well designs can improve the conduction- and valence-band overlaps by modifying the overall amount of band bending in the quantum well. However, the device design and the fabrication of devices becomes that much more complicated.

Nanowire LEDs and Lasers

Epitaxial nanowires are attractive for relieving strain, reducing dislocation densities for the same nominal lattice mismatch, and for reducing polarization effects. This can lead to higher gains, lower threshold current densities, and faster modulation. The growth conditions need to be optimized to obtain the right compositions and strains. Figure 5.5 shows typical group III-Nitride quantum-wire LED characteristics.

Nanowire White LEDs

Core-shell nanowires lead to significantly reduced built-in polarization fields and better quantum efficiencies for LEDs. Further, they can lead to broad spectrum emissions and the realization of intrinsic white LEDs. Figure 5.6 shows the emission spectrum from a specific core-shell nanowire LED structure. Growth optimization can help achieve control over the color rendering of the white LED. Efficient strain relaxation is one of the advantages offered by nanowire LEDs and can lead to improved quantum efficiencies and optical power output for given current densities.

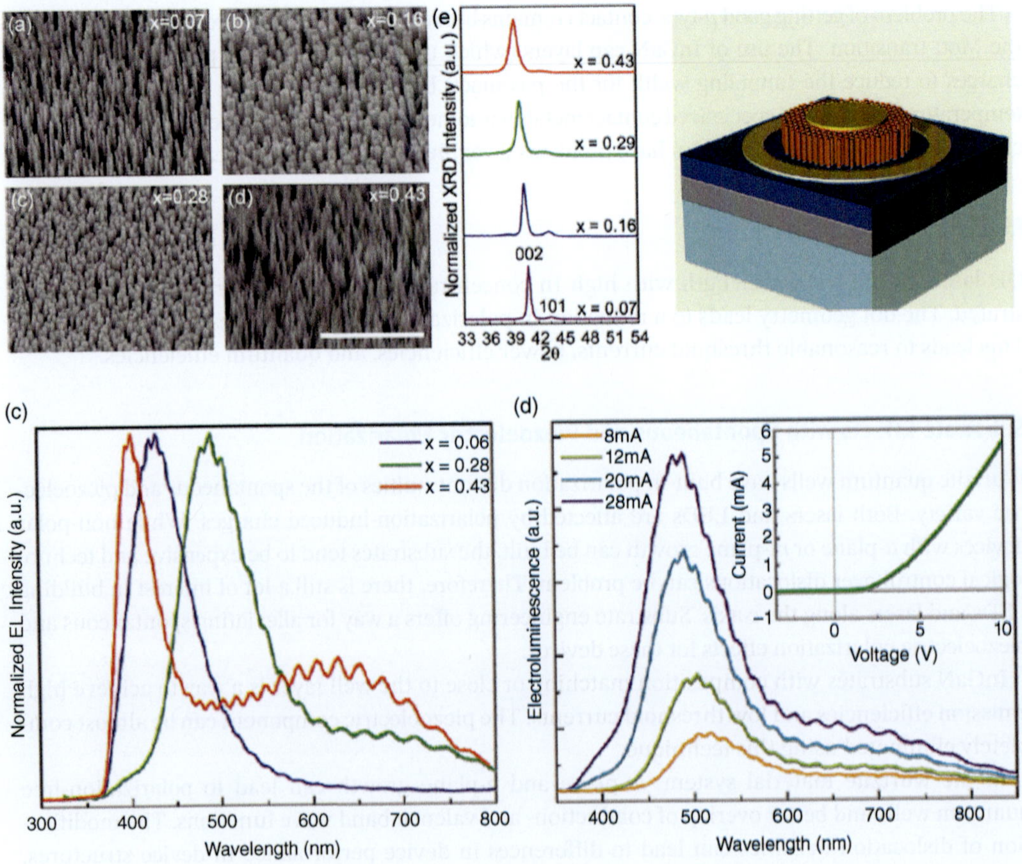

Figure 5.5 (a) Scanning electron microscopy (SEM) images of InGaN nanowires. (b) A schematic of an LED device. (c,d) InGaN LED electroluminescence intensity for different In composition and electroluminescence at In composition 0.28 for various currents – from Reference [2], used with permission.

Nanowire microLEDs

Nanowire microLEDs in the InGaN system are of great interest because they can give better carrier confinement, better electron-hole overlap, and better overall efficiencies compared to conventional microLEDs. InGaN/AlGaN nanowire microLEDs with good efficiency have been demonstrated in Reference [3]. The nanowire LEDs consist of GaN:Si grown on a silicon substrate and the GaN:Mg on the top. The 10 couples of quantum wells are inserted in the active region. Each quantum well includes a 3 nm InGaN dot and 3 nm AlGaN barrier. During growth of the AlGaN barrier an AlGaN shell is formed with unique core-shell layers. The nanowire LED characteristics are shown in Figure 5.7.

LED Quantum Efficiencies and Laser Threshold Currents in Quantum Wires and Quantum Dots

Quantum wires and QDs have severely reduced density of states compared to quantum wells. For the same bias level in *n-i-p*-type LEDs this means there are fewer conduction-band electrons and valence-band holes at the same bias level. This leads to a lower intensity of LED emission at

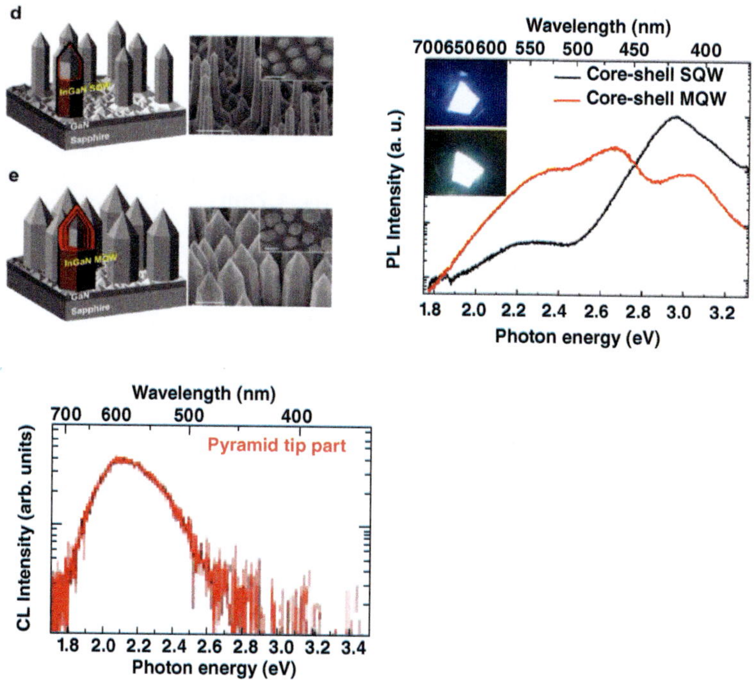

Figure 5.6 Nanowire white LED SEM images (top left), photoluminescence spectrum (top right), and cathodoluminescence (bottom left) – from Reference [2], used with permission.

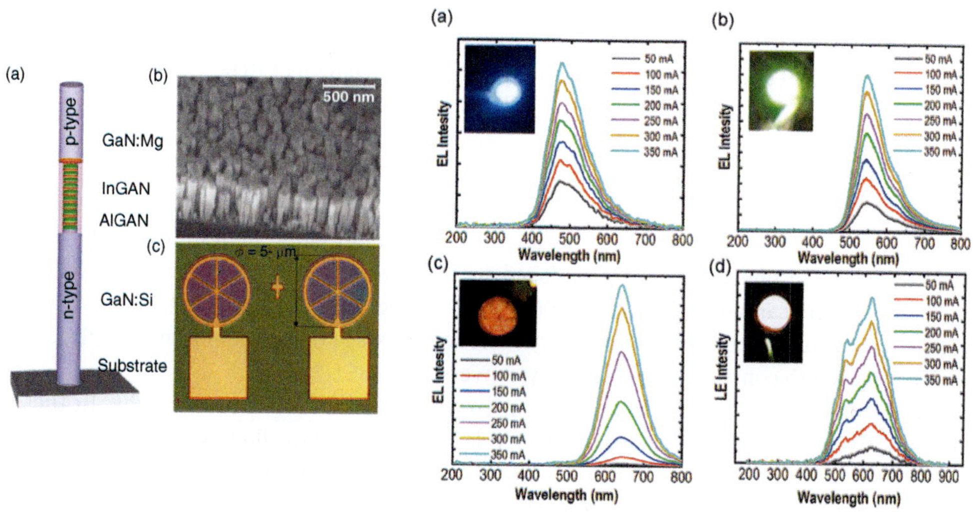

Figure 5.7 InGaN/AlGaN nanowire LED structure (left) and emission characteristics of blue, green, red, and white microLEDs (right). From Reference [3], used with permission.

comparable bias levels when there are no dislocations. However, quantum wires and dots tend to have lower dislocation densities due to the ability to relax lattice mismatch-induced strain more easily. Therefore, it is possible to obtain better quantum efficiencies by reducing the non-radiative recombination component compared to the radiative recombination component.

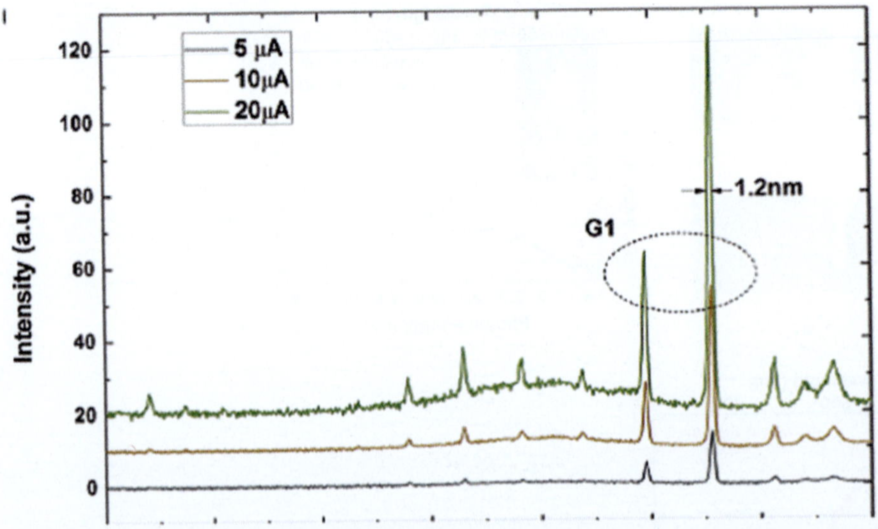

Figure 5.8 Laser emission spectrum at different currents. After Reference [104], used with permission.

Quantum Dot Lasers

Wurtzite nitride QD layers have been used to obtain red lasers with low threshold currents. The cavity dimension design is of crucial importance for these devices.

Quantum Dot LEDs

QD LEDs can alleviate the droop problem, for instance, in the green emission region due to lower Auger rates. Figures 5.9 and 5.10 show typical outputs of optimized QD LEDs.

Quantum Dot-based White LEDs

The conventional white LED technology uses blue or UV LEDs along with a phosphor for conversion into other higher wavelengths so as to obtain overall white light emission. In more advanced designs, the phosphor can be replaced by QD layers to act as frequency conversion layers. The QD layers can absorb the blue or UV light and get excited, followed by relaxation to lower excited states and then to the ground state. The second stage of emission from the QD leads to effective wavelength conversion or frequency conversion.

Multiple QD layers are often needed to obtain conversion to different wavelengths. In these designs, the QDs in different layers have energy levels that are tuned to allow conversion to different wavelengths of light. Sometimes, some of the QD layers are built from other material systems, that is, material systems different from group III-Nitride semiconductors. This is especially so for obtaining conversion to the red end of the spectrum. The reader can find details of group III-Nitride nanowire and QD LEDs and lasers in References [4–42].

Auger Recombination and Efficiency Droop in Group III-Nitride LEDs

Several reports on group III-Nitride LEDs, specifically InGaN LEDs, show that the quantum efficiency of the LEDs drops at higher current density. Further, this efficiency droop becomes more serious as the wavelength of the optical emission is increased. Therefore, this phenomenon

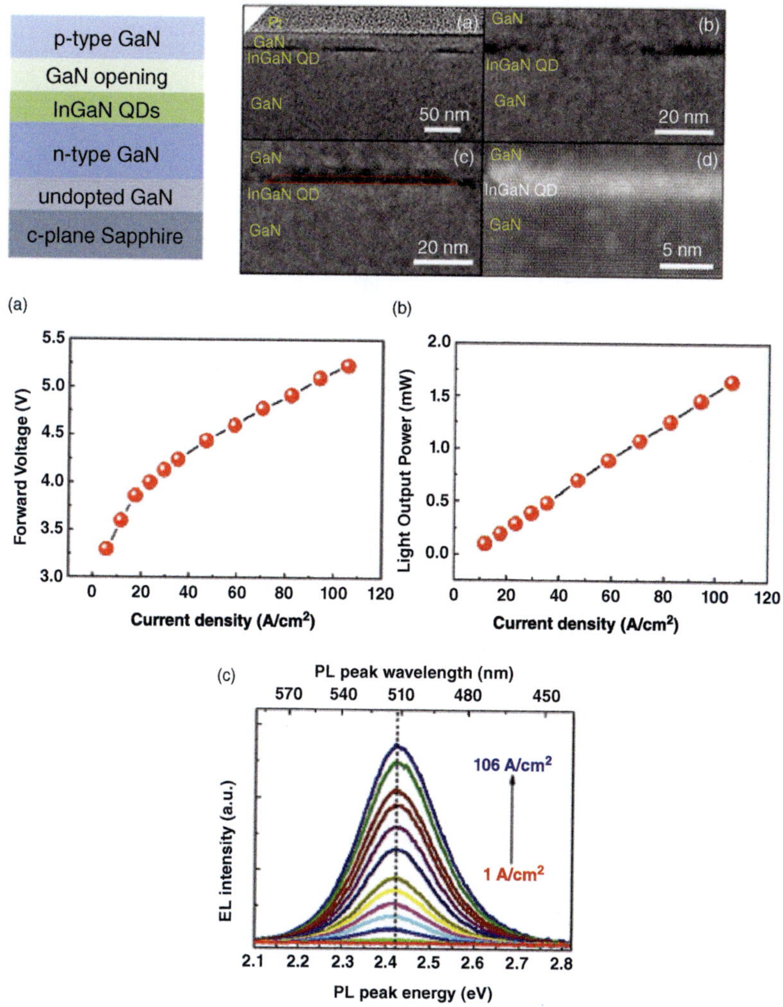

Figure 5.9 (a) Transmission electron microscopy (TEM) image of InGaN quantum dot (QD) LED. (b) QD LED output characteristics – after Reference [43], used with permission.

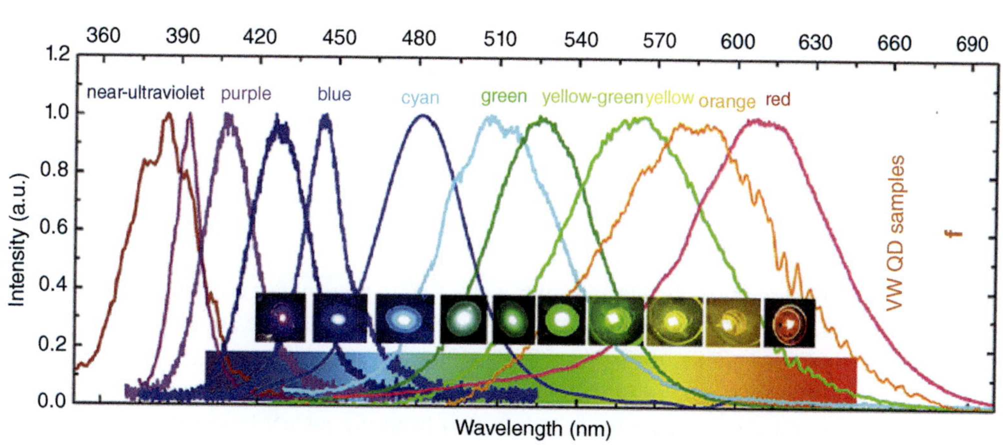

Figure 5.10 Wavelength-tunable LED output of QD microLEDs – after Reference [44], used with permission.

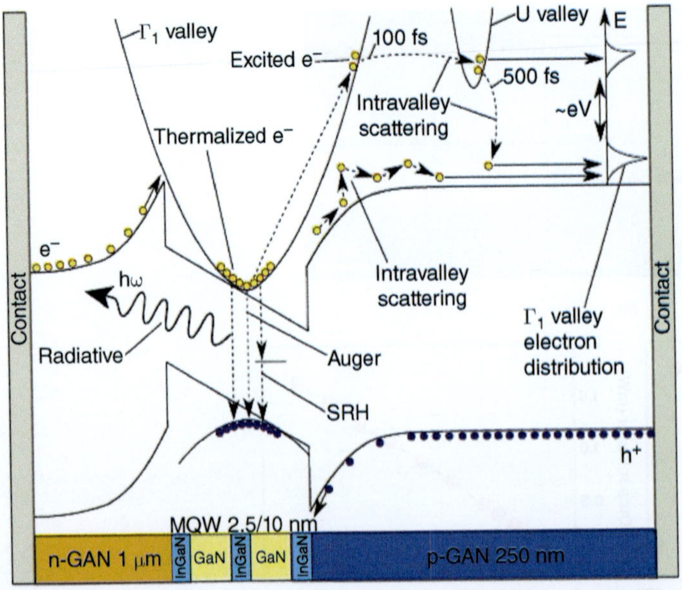

Figure 5.11 Various scattering and recombination mechanisms in a quantum-well InGaN LED – after Reference [45].

is also called the "green gap" problem. Several experimental and theoretical investigations have found that this efficiency droop can be traced to higher rates of non-radiative Auger recombination at high bias levels.

Figure 5.11 shows various microscopic scattering and recombination mechanisms in an LED. Reference [45] uses the Monte Carlo method combined with drift-diffusion transport to study the cause of efficiency droop in InGaN LEDs. They find that the hot electron population created by Auger recombination in the quantum wells (QWs) is trapped in the high-energy side valleys and transported to the semiconductor surface through a *p*-type barrier layer. They identify Auger recombination as the major cause of electron emission over the barrier, which exacerbates the efficiency decrease of the LED. The simulations in Reference [45] indicate that an EBL cannot be too effective in reducing efficiency droop in InGaN LEDs.

Non-polar plane LEDs are better at reducing the efficiency droop because they have better overlap of electron and hole wave functions, and therefore, they have higher radiative recombination rates. This compensates for the Auger recombination and provides overall higher efficiencies compared to *c*-plane LEDs. Hybrid quantum-well structures with higher In-content InGaN wells near the *p*-GaN on top and lower In-content wells near the *n*-GaN on the bottom. Reference [46] shows that this hybrid design decreases dislocation densities and lowers non-radiative recombination, which leads to higher efficiencies.

Electron-blocking Layer and Effect on Hole Transport

EBLs are often used in group III-Nitride quantum-well LEDs to prevent fast escape of hot electrons from the well layer. The EBL in a GaN/InGaN LED typically consists of an ultra-thin layer of *p*-AlGaN with a bandgap significantly higher than GaN. The hot electrons injected from the *n*-layer into the quantum well cannot escape into the *p*-layer due to the tunneling barrier provided by the EBL. The EBL also provides a barrier for hole injection from the *p*-GaN layer into the quantum well. This leads to lower population inversion at given bias levels. The Mg acceptors in the AlGaN

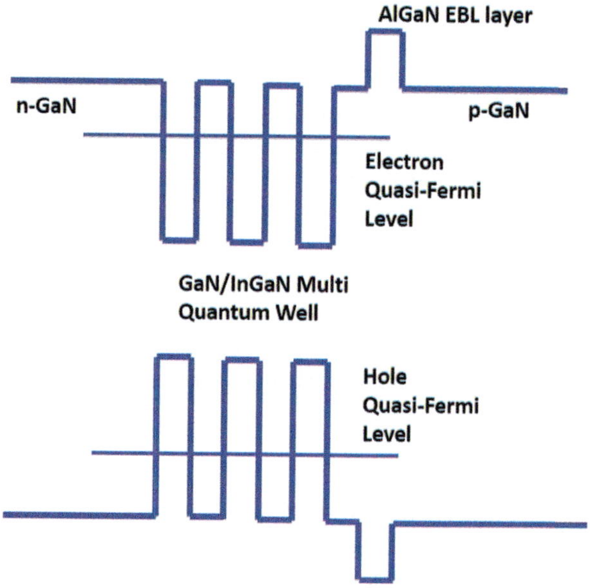

Figure 5.12 Schematic diagram of an electron-blocking layer (EBL) in an InGaN emitter.

layer tend to have a low ionization ratio due to large hole masses and so they cannot help eliminate this problem. Figure 5.12 shows typical band diagrams with an EBL layer.

To obtain EBLs that do not lead to the degradation of hole injection efficiencies, superlattice EBLs have been explored in the literature. Superlattice *p*-layers lead to higher hole concentrations in the superlattice due to band-bending effects and due to the delocalization of holes from the higher bandgap AlGaN layers to the lower bandgap GaN layers. This also leads to a higher injection efficiency from the *p*-type cladding layer into the quantum well. Superlattice layers come with added design complexity and added process complexity, however. Therefore, they can be used in cases where the performance improvement makes them worthwhile.

The EBL also has polarization-induced charges at both interfaces. This can reduce the effective barrier for electron escape in many situations. Using graded AlGaN EBLs sometimes improves EBL performance in some designs. The graded EBL typically uses *p*-AlGaN graded from AlN to GaN over part of the EBL and *p*-AlGaN of fixed composition for the rest of the EBL. This design increases the effective barrier seen by the quantum-well electrons and therefore blocks electrons effectively.

AlGaN Barriers as Alternative to GaN Barriers in InGaN LEDs and Lasers

AlGaN barriers can be expected to have better electron-blocking capability because the well-barrier conduction-band discontinuity is intrinsically higher. However, it must be kept in mind that for the same well composition AlGaN barriers can lead to a higher quantum-confined Stark effect. This means a lot of optimization is needed to obtain the right kind of AlGaN barrier. The use of strained versus unstrained AlGaN barriers needs to be compared numerically before arriving at the optimum LED or laser design.

AlGaN barriers, however, lead to problems with the activation of acceptors and donors in the barrier layers. Careful optimization and specialized processing methods as well as specialized device tricks like superlattice and resonant doping can become necessary to get workable devices.

GaN/InAlGaN/InGaN/InAlGaN/GaN and GaN/InAlN/InGaN/InAlN/GaN structures have been explored as alternatives to reach the right balance between polarization effects and electron-blocking and carrier-confinement needs. The use of quaternary alloys like InAlGaN as quantum marriers provides extra design dimensions in terms of lattice constant, lattice mismatch, critical thicknesses, and spontaneous and piezoelectric polarization. However, the right composition control of quaternary InAlGaN is much more difficult technologically.

Electron-cooling Layer to Improve Efficiency

An InGaN layer inside the *n*-GaN barrier layer can be used as an electron-cooling layer (ECL) to prevent loss of electrons from the well to the *p*-layer. The ECL adds process complexity and also requires careful optimization of polarization charges to obtain desired characteristics.

Defect-state Saturation and Radiative Recombination Efficiency

Mismatched heteroepitaxy can lead to significant dislocation density in the active well layer. This can happen despite efforts to keep the well-layer thickness below critical thickness because the actual critical thickness is a complicated function not just of lattice mismatch but also of growth conditions and thermal mismatch. Carrier-induced defect saturation leads to several of the defective mid-gap states being inactive for non-radiative recombination. Therefore, the radiative recombination efficiency can depend on bias levels and needs to be carefully calibrated in actual devices.

Quantum Barrier Engineering in InGaN Emitters to Alleviate Polarization Effects

Si-doped quantum barriers can be used in the quantum-well region to enhance electron concentration so that the resulting screening effect can reduce the quantum-confined Stark effect. However, trade-offs with hole injection efficiency must be contended with when these designs are used. S-doping sometimes leads to improved interfacial quality of multi quantum-well layers.

Dislocations in Low-dimensional Structures and Carrier Dynamics

Critical Thicknesses and Dislocation Formation

The lattice mismatch between InN and GaN is large and even low indium content InGaN can have a large percentage lattice mismatch. The critical thickness for the formation of misfit dislocations as a function of indium concentration on GaN can be calculated according to models from the literature, like the Matthews–Blakeslee and the Bean–People models.

It must be kept in mind that the actual critical thickness can depend on kinetic factors during non-equilibrium growth and often needs to be verified empirically. The actual critical thicknesses can be different for wells, wires, and dots again due to kinetic considerations.

Recombination Dynamics of Carriers

A key feature of light-emitting as well as energy conversion devices is the carrier recombination dynamics in the presence of dislocation-related or other type of traps. The overall recombination rate is given by

$$R = An + Bn^2 + Cn^3 \tag{5.39}$$

where the An term is due to Shockley–Read–Hall-type recombination via mid-gap states, the Bn^2 term is due to direct recombination of conduction-band electrons and valence-band holes, and the Cn^3 term is due to Auger recombination in which the electron recombines by giving energy to another conduction-band electron.

Disorder, Carrier Localization, and Effect on Recombination and Red Shifts

Ab Initio Density Functional Theory Calculations

Since InGaN alloys have alloy disorder, the Bloch states of periodic crystals can become drastically modified. Most of the carrier effective mass determinations using techniques like electron cyclotron resonance and infrared ellipsometry have been done on non-alloyed GaN and InN. Theoretical work points to the bowing of InGaN effective mass as the composition is changed from GaN to InN.

Localization and Light Emission

A key feature of light-emitting and energy conversion devices is the carrier recombination dynamics in the presence of dislocation-related or other type of traps. The recombination rate due to traps is given by a well-known Shockley–Read–Hall type of expression and must be added to direct inter-band recombination and Auger recombination to get the total rate. There is evidence from electronic structure calculations that the valence-band wave functions can be highly localized when the indium content is high in InGaN. This means they are no longer delocalized Bloch-type wave functions, and therefore, the effective mass approximation cannot be used to calculate states in quantum wells, quantum wires, and QDs. The localization seen in the bulk InGaN is likely to persist in wells, wires, and dots. That means the quantum mechanical transition rates between conduction-band states and valence-band states will be dominated by this localization, and for high levels of localization the transition rates can be severely reduced. Therefore, the indium content cannot be too high in InGaN/GaN optoelectronic devices if good efficiency is to be obtained.

Staggered Quantum-well InGaN Laser Characteristics

The use of staggered configurations alleviates the band-bending problem inside the quantum wells. The quantum-well composition undergoes a step shift in these designs, and a higher indium composition layer is followed by a lower indium composition layer. Since the polarization discontinuity at the lower indium/higher indium layers interface has the opposite sign of that at the GaN/InGaN interface looking downwards from the top, the electric field undergoes a step shift of the opposite sign. This is the origin of the lower band bending in these staggered quantum wells.

$$\sigma_1 = P_{1_{\text{spontaneous}}} + P_{1_{\text{piezoelectric}}} - P_{2_{\text{spontaneous}}} - P_{2_{\text{piezoelectric}}} \tag{5.40}$$

$$\sigma_2 = P_{2_{\text{spontaneous}}} + P_{2_{\text{piezoelectric}}} - P_{3_{\text{spontaneous}}} - P_{3_{\text{piezoelectric}}} \tag{5.41}$$

The electric field is given by the continuity of the electric displacement vector at the interface in the absence of free interface sheet charge

$$D_1 = D_2 \tag{5.42}$$

$$\varepsilon_1 E_1 + P_{1_{\text{spontaneous}}} + P_{1_{\text{piezoelectric}}} = \varepsilon_2 E_2 + P_{2_{\text{spontaneous}}} + P_{2_{\text{piezoelectric}}} \tag{5.43}$$

Similar equations hold for layers two and three as well. The magnitude of the electric field discontinuity at typical *c*-axis GaN/InGaN heterojunctions is on the order of $10^6\,\text{V}\,\text{cm}^{-1}$. Since in staggered-well designs σ_2 have opposite signs, the total band bending is less than in a single quantum-well design of same width. Figures 5.13–5.16 show schematics of band diagrams for nonstaggered- and staggered-well designs as well as their gain spectra for a specific set of InGaN quantum well-laser designs.

In staggered-well InGaN/GaN designs, the quantum-well layer is basically split into two layers with different In content. The overall well width cannot be increased too much due to this though because the effect of decreased band bending will be lost if this is done. This means that

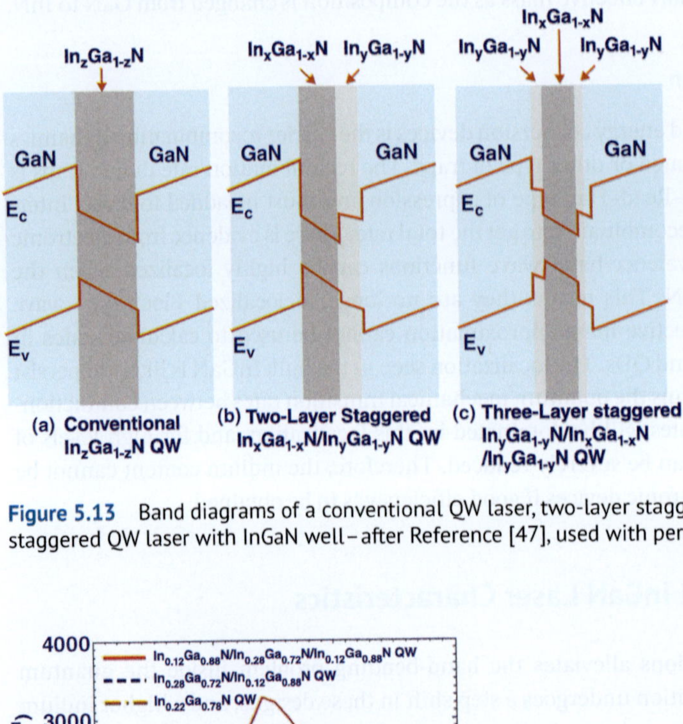

Figure 5.13 Band diagrams of a conventional QW laser, two-layer staggered QW laser, and three-layer staggered QW laser with InGaN well – after Reference [47], used with permission.

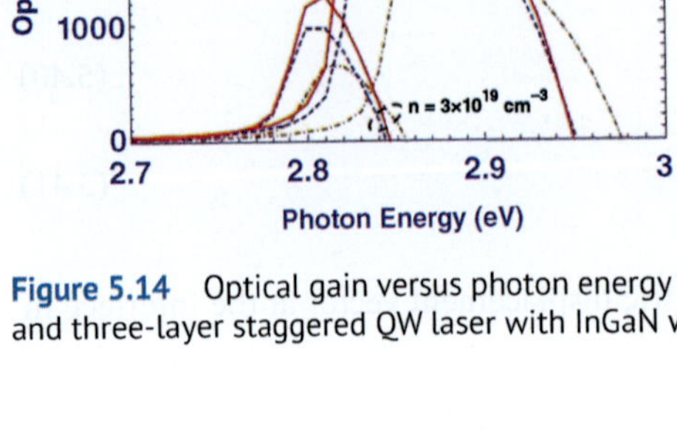

Figure 5.14 Optical gain versus photon energy for a conventional QW laser, two-layer staggered QW laser, and three-layer staggered QW laser with InGaN well – after Reference [47], used with permission.

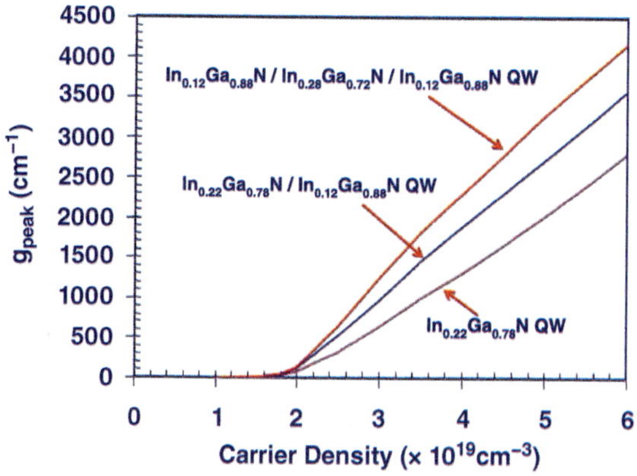

Figure 5.15 Peak gain versus carrier density for a conventional QW laser, two-layer staggered QW laser, and three-layer staggered QW laser with InGaN well–after Reference [47], used with permission.

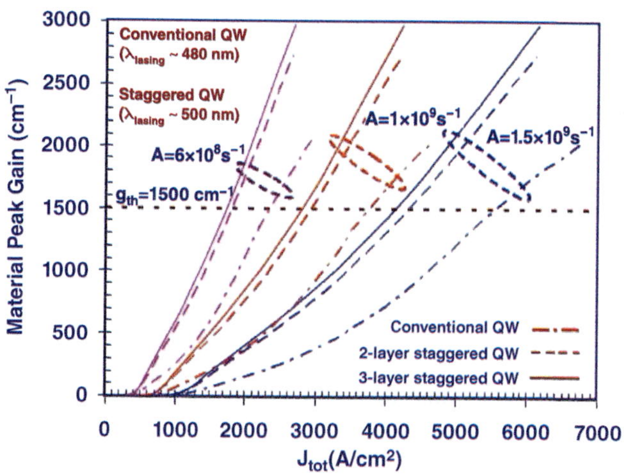

Figure 5.16 Material peak gain versus current density for conventional QW laser, two-layer staggered QW laser, and three-layer staggered QW laser with InGaN well–after Reference [47], used with permission.

staggered-well designs need more precise control over epitaxial growth compared to nonstaggered-well designs. This kind of control is typically easier to achieve with MBE compared to MOCVD, but it comes with all the added complications due to the ultra-high vacuum requirements of MBE.

Non-polar Plane Quantum-well InGaN LEDs and Lasers

While non-polar plane InGaN growth (*a*-plane growth or *m*-plane growth) is more expensive, it can completely eliminate the built-in electric fields along the growth direction. This leads to better overlap of electron and hole wave functions, higher gain for the same bias levels, better optical output for the same current density, and faster modulation. Figure 5.17 shows the modulation parameters of a non-polar plane LED from InGaN. Non-polar plane growth (*a*-plane or *m*-plane growth) is more involved technologically compared to polar growth and is therefore more expensive.

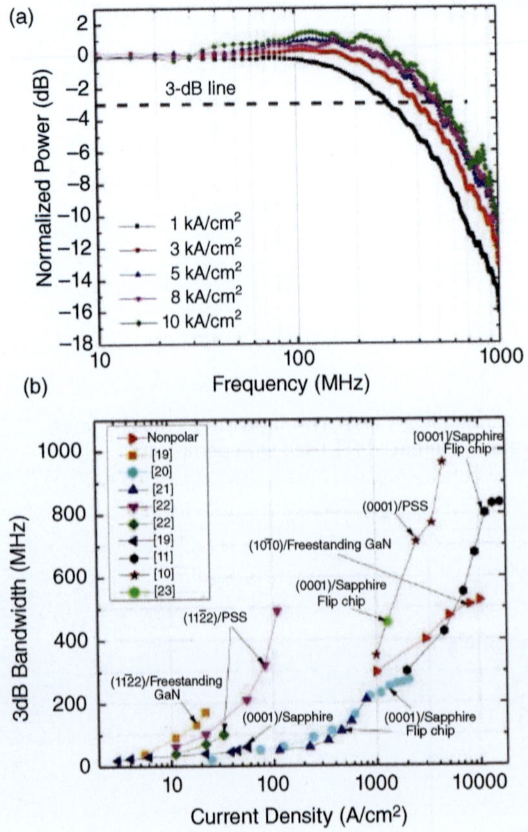

Figure 5.17 Modulation parameters of InGaN laser – after Reference [48], used with permission.

Continuous improvements in technological capabilities may make non-polar plane growth competitive with polar plane growth in the future. There are also semi-polar plane group III-Nitride layers, which have polarization magnitudes intermediate between polar and non-polar planes. These semi-polar plane devices can also benefit from the right kind of staggered-well designs.

Semi-polar Plane Quantum-well InGaN LEDs and Lasers

Semi-polar and non-polar plane growth on typical substrates, like sapphire, often leads to high defect densities like basal stacking faults and threading dislocations. This leads to low quantum efficiencies and low power efficiencies as well. Therefore, a lot of the work on semi-polar and non-polar GaN/InGaN quantum-well devices have focused on GaN substrates, which are expensive. Semi-polar plane growth can sometimes give lower defect densities in device layers compared to non-polar plane growth, and therefore, semi-polar plane LEDs and lasers are of interest.

Novel growth methods have been explored for semi-polar and non-polar plane growth on sapphire, SiC, and Si substrates. One method is the growth of GaN template layers on patterned sapphire [49]. Semi-polar sapphire is patterned to reveal (0001) facets. GaN layers grow on the 0001 facets and the islands of GaN coalesce to provide semi-polar face GaN. The use of GaN templates leads to lower dislocation densities in the semi-polar device layers and helps achieve higher device efficiencies.

The substrate technology for group III nitrides is constantly evolving and improving. The costs are decreasing as more research leads to improved layers and growth processes. Therefore, semi-polar growth directly on semi-polar GaN substrates may become competitive for some devices and some applications with time.

p-Type Ohmic Contacts and Efficiency of LEDs and Lasers

Ohmic contacts to the *p*-type cladding layer of lasers and LEDs is a major technological challenge if one wants to improve the power efficiency of these devices. Also, the series resistance of the *p*-type layer can become a device performance limiter unless properly optimized. Magnesium (Mg) is the most common dopant in *p*-type GaN or *p*-type AlGaN, and it has several physical parameters that make it difficult to get high *p*-type conductivity. Mg is a deep acceptor in GaN and AlGaN due to the large hole effective mass in these materials and the low high-frequency dielectric constant. The hole effective masses in GaN and AlGaN are in the range of 1–1.5 times the free electron mass. The high-frequency dielectric constant is in the range of five to six. Together, these lead to observed acceptor ionization energies of 200–250 meV. This is 8–10 times larger than the room-temperature thermal energy of 26 meV. Figure 5.18 shows the location of the Fermi level in *p*-type GaN for two different physical doping densities. The distance of the Fermi level from the valence-band edge is much larger than typical values observed in lower bandgap *m* materials like AlGaAs or GaAs. This leads to ionization ratios of 0.1–0.15, which means that getting a $10^{19}\,cm^{-3}$ hole carrier concentration requires a physical doping density of $10^{20}\,cm^{-3}$ or higher. This kind of high doping density along with low ionization ratios leads to very high scattering rates from non-ionized impurities and hole mobilities can be as low as 1–2 $cm^2V^{-1}sec^{-1}$.

In addition to the high *p*-layer resistance, this relatively low hole carrier concentration also contributes to increased *p*-type ohmic contact resistance. The large hole effective masses mean that the tunneling transmission coefficients across the ohmic contact barrier is exponentially suppressed in these layers. The Wentzel–Kramers–Brillouin (WKB) transmission probability across a layer with width W and barrier height Φ_b is given by

$$T = exp\left(-2\int dx\sqrt{\frac{2m^*}{\hbar^2}}\sqrt{(V(x)-E)}\right)$$

Typical Schottky barrier heights to GaN or AlGaN tend to be 1 eV or larger, compared to much lower values in GaAs or Si. Therefore, the $\sqrt{(V(x)-E)}$ factor tends to be large. m^* is also relatively large, as already mentioned. The width of the barrier tends to be large because of the large barrier

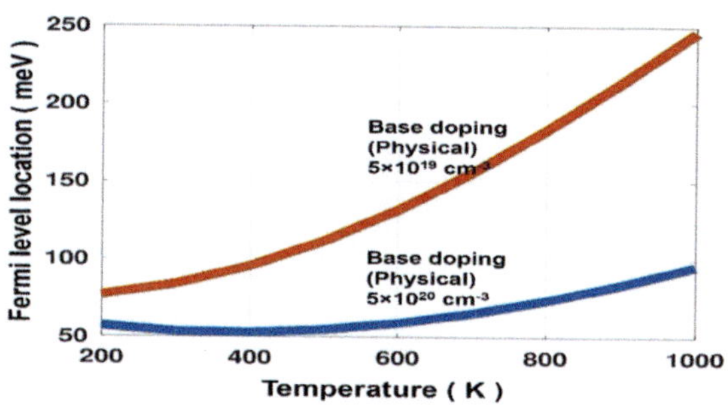

Figure 5.18 Location of hole Fermi level measured from the top of the valence-band edge for different physical Mg doping densities in GaN.

heights, as can be derived from the depletion layer equation at a tunneling Schottky barrier. That means the integral inside the exponential extends over a much larger region. Together, these three factors lead to exponential suppression of the transmission probability at any given hole energy. The ohmic contacts obtained through non-optimized contacts tend have specific contact resistances that are too large for getting good device power efficiencies.

One way to reduce the contact resistance is to use specialized contact metals that give lower Schottky barrier heights after annealing. Typically, very-high temperature anneals in inert environments are required to obtain low Schottky barrier heights. This can pose a problem for process integration in this material system because GaN tends to have a low sublimation temperature. A room-temperature method to obtain low specific contact resistance is therefore preferred from the device integration perspective. One way to obtain lower specific contact resistance without high-temperature anneals is to use polarization engineering. Typically, using ultra-thin (3 nm or thinner) InGaN cap layers on GaN can reduce the barrier width by a factor of 2.5–3. This is because there is a negative charge at the InGaN/GaN interface due to polarization discontinuity. This polarization-induced interface sheet charge leads to a much lower barrier width, which leads to tunneling transmission probabilities that are four to five orders of magnitude probabilities larger than in non-optimized contacts.

It must be noted that the polarization method can only be used for *c*-axis growth because that is the polar axis. However, as already discussed earlier in this chapter, the *c*-axis growth leads to lower momentum matrix elements and lower gains unless some special tricks like staggered quantum-well designs are used. Therefore, the staggered quantum-well design must be used in combination with thin InGaN caps to solve the *p*-type contact resistance problem. There is a lot of interest in semi-polar and non-polar axis LEDs and lasers because of the advantage they offer in terms of higher intrinsic radiative recombination rates and higher intrinsic gains. The polarization-based method with ultra-thin InGaN cap layers cannot be used for these devices because the polarization is low or there is no polarization along the growth direction. In this case, special materials optimization methods like annealing with electron irradiation become necessary.

Superlattice Structures to Increase *p*-Type Barrier Layer Conductivity

Superlattices of AlGaN/GaN and superlattices of GaN/InGaN can be used in the *p*-type barrier layers of UV LEDs and lasers as well as visible light LEDs and lasers, respectively. Superlattices form minibands that can become resonant with the Mg acceptor level, which leads to better ionization and enhanced *p*-type conductivity. Figure 5.19 shows the typical power output improvement in LED with superlattice *p*-layer.

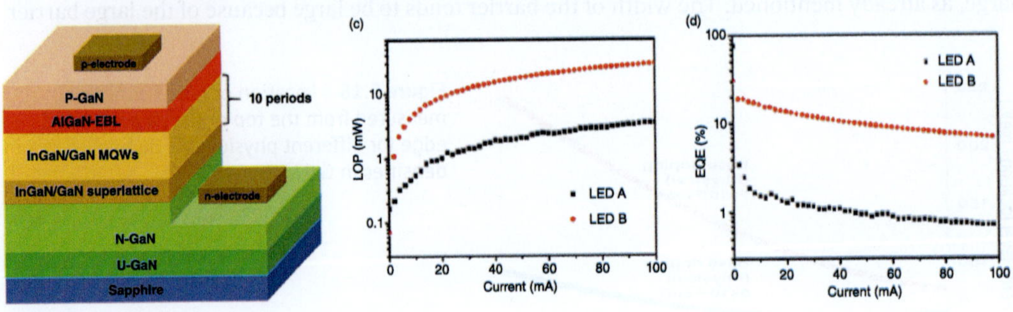

Figure 5.19 Power output improvement in LED with superlattice *p*-type layer. From Reference [50], used with permission.

Vertical Cavity Surface Emitting Lasers

Vertical cavity surface emitting lasers (VCSELs) can be fabricated in the InGaN material system with a vertical stack of distributed Bragg reflector (DBR) layers. The reader is referred to standard textbooks on semiconductor optoelectronics and the literature for basic operation of VCSELs. In the InGaN material system, the design has to take into account the usual changes in material parameters.

VCSELs can offer higher modulation speeds with proper design. They also have potential applications in solid-state lighting, high resolution printing, and biochemical sensing. Reference [51] demonstrated continuous wave room-temperature operation of a VCSEL in the "green gap" regime at a yellow-green wavelength of 560.4 nm. The VCSEL demonstrated there are InGaN/GaN QD layers as the active layers. The DBR layers were fabricated using TiO2/SiO2 multilayers. The active layers and cladding layers were grown on a sapphire substrate instead of the more expensive GaN substrate. The density of dislocations and non-radiative dislocation centers were kept low using a QD active layer since the Stranski–Krastanow QDs tend to have low dislocation densities.

The QD strain distribution has a lower strain magnitude compared to quantum-well designs and the piezoelectric field is reduced considerably. The laser was bonded upside down to a highly conductive copper (Cu) substrate to reduce steady-state temperature during continuous wave operation. Figure 5.20 shows characteristics of a typical InGaN QD VCSEL.

Microcavity VCSEL-type laser operation was demonstrated with a normally blue emitting layer in a microcavity confinement structure to obtain green wavelength emission under high injection (Reference [52]). Wavelength-tuneable UV-VCSEL with the wavelength range 376–409 nm was demonstrated in Reference [53]. Blue and green GaN-based VCSELs with epitaxial AlInN/GaN DBR layers were demonstrated in Reference [54]. Epitaxial DBR structure offers scope for easier process integration and can potentially reduce cost and manufacturing complexity.

High quality AlN/GaN DBRs are a good alternative to obtain high-efficiency GaN VCSELs. Epitaxial growth optimization and co-design with DBR optical characteristics using layer dielectric constants and thicknesses are necessary to obtain required performance from these VCSELs.

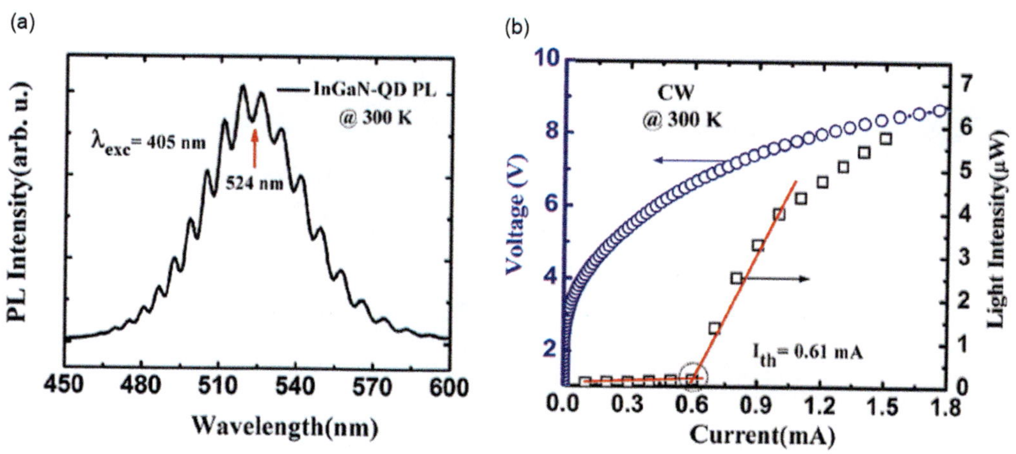

Figure 5.20 Photoluminescence of InGaN QD layer (left) and laser output characteristics (right) of InGaN QD VCSEL. After Reference [51], used with permission.

Distributed Feedback Lasers

Distributed feedback (DFB) lasers in the InGaN material system use a grating in the lateral direction to obtain high wavelength selectivity of reflectivity. This helps obtain a very low linewidth or extremely monochromatic lasing in the blue wavelength for potential applications in metrology and spectroscopy. Single mode and low linewidth lasing is of potential use in optical atomic clocks, high-precision spectroscopy, biomedical applications, visible light communications, and light detection and ranging (LiDAR) sensing and surveying. Typical structure and emission characteristics of a DFB laser are shown in Figures 5.21 and 5.22. The gratings for distributed feedback can be realized as surface gratings, sidewall gratings, or buried heterostructure gratings. Preliminary results of low linewidth lasing from DFB InGaN/GaN laser on Si substrates have been reported in Reference [55]. The reader can find details of various DFB lasers in References [56–65].

Plasmonic Nanolasers

The use of plasmonic dispersion can help achieve lasing operation at much lower wavelengths compared to the free-space wavelength. Therefore, the dimensions of the lasers can be in the nanometer range, thereby achieving nanolaser operation. The plasmonic nanolasers, also called SPASER (surface plasmon amplification by stimulated emission of radiation), help overcome optical diffraction limits and reduce laser size. Several material and design challenges remain to fully optimize these nanolasers in the InGaN system. The electric field profile in the SPASER is crucial to determining the threshold current density of the laser because good overlap of the electric field profile with the exciton wave function in a nanorod gain-providing layer, for example, is needed to obtain high gain coefficients.

SPASER operation with InGaN/GaN nanorods separated from the silver (Ag) layer by SiO_2 has been simulated in Reference [66]. Spontaneous emission enhancement due to the Purcell effect is observed, and the simulations in the reference provide a path for SPASER fabrication and optimization.

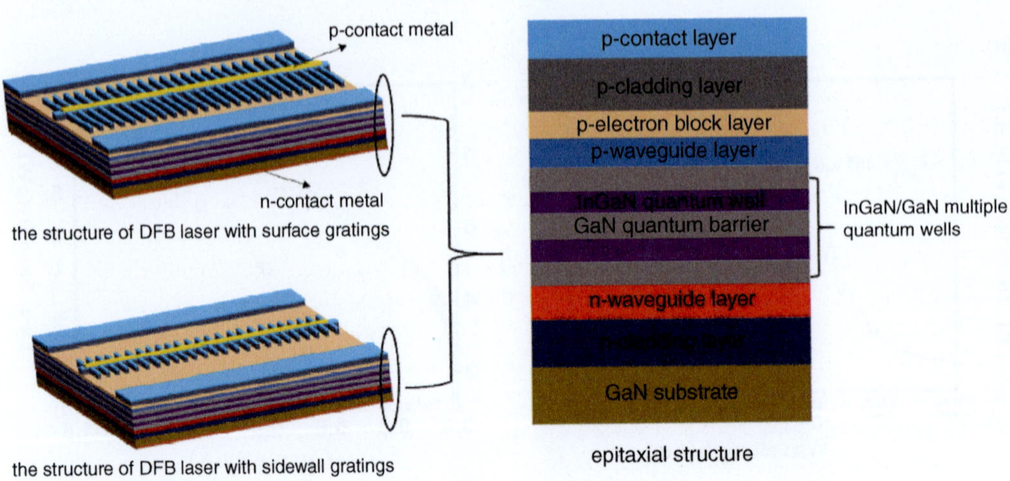

Figure 5.21 Typical structure of distributed feedback laser. From Reference [67], used with permission.

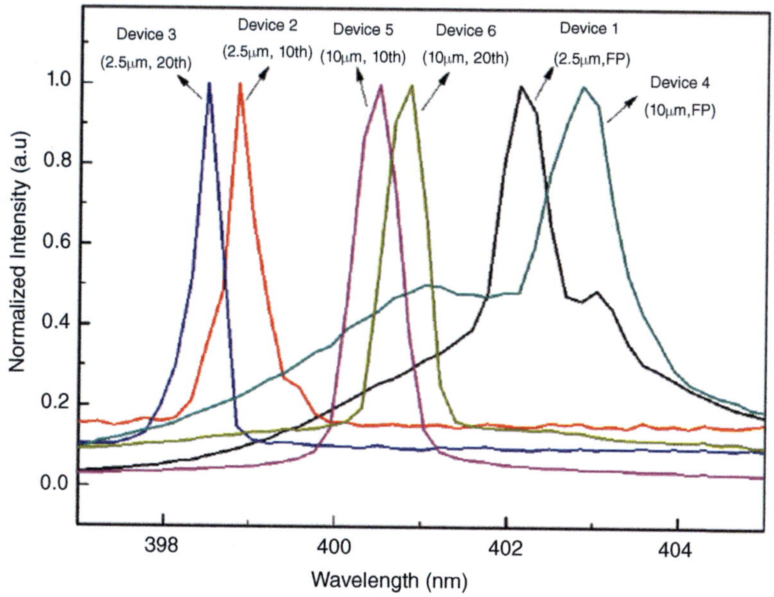

Figure 5.22 Narrow linewidth emission from DFB InGaN/GaN lasers. From Reference [67], used with permission.

Surface Plasmon-enhanced LEDs

Surface plasmon-enhanced LEDs use the plasmon coupling to attain enhancement of spontaneous emission rates and higher radiative efficiencies. Reference [66] demonstrates enhancement effects in multi quantum-well InGaN LEDs with silver (Ag) nanoparticles that provide the surface plasmon coupling of the electric field modes of the LED. Enhancement factors approaching 100 have been demonstrated, as shown in Figure 5.23.

Indium Gallium Nitride LEDs and Lasers on Si Substrates

Typically, (0001) InGaN can be grown on (111) Si with significant strain. The GaN barrier layers of InGaN LEDs and lasers have an even higher strain on Si compared to InGaN. The use of a pre-strain layer of InGaN before the InGaN quantum-well layers can help alleviate strain effects and dislocation densities substantially.

GaN/InGaN template layers for the buffer and for the barrier layers can lead to significant strain relief and also lower dislocations densities in the GaN barrier. Propagation of dislocations from the GaN barrier to the InGaN quantum-well layers can then be reduced. The template method can, therefore, lead to improved internal and external quantum efficiencies.

UV lasers and LEDs typically rely on AlGaN and GaN layers, both of which have a large lattice mismatch, almost 15%. This means that both the barrier layers and the quantum-well layers have intrinsic dislocations. The use of template buffers and barrier layers is, therefore, even more beneficial for UV lasers and LEDs on Si substrates. InN has a much lower lattice mismatch with Si. Therefore, InGaN layers should be incorporated in AlGaN and GaN layers to reduce effective mismatch and strain. The use of quaternary quantum barriers and corresponding polarization engineering can also potentially yield benefits, though good optimization is needed to attain device characteristics goals like quantum efficiencies and power efficiencies.

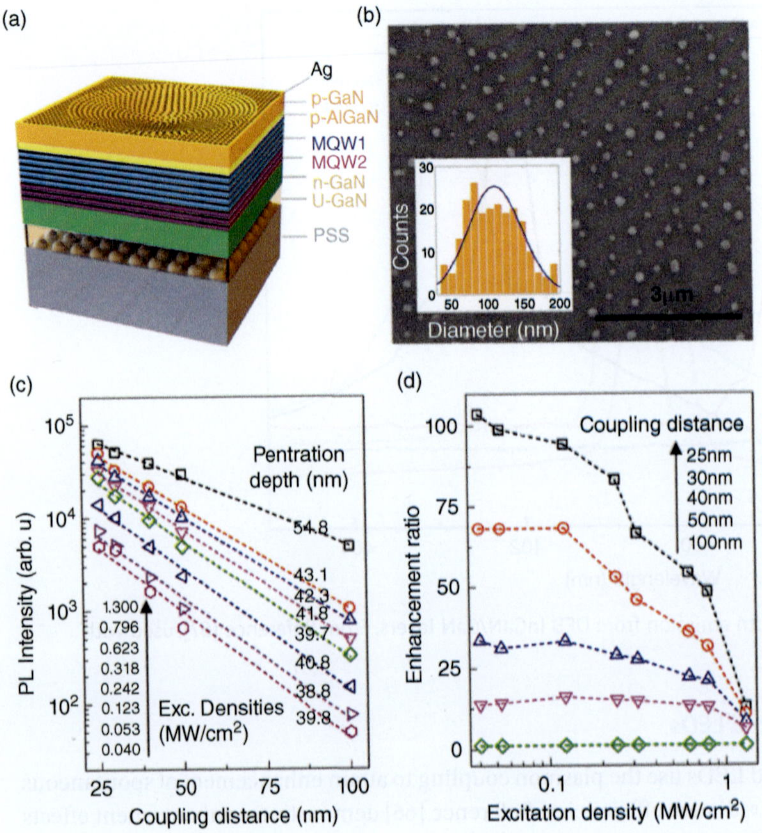

Figure 5.23 Surface plasmon-enhanced efficiency in InGaN light emitters. After Reference [66], used with permission.

Vertical Versus Lateral LEDs

Vertical LED structures can be created by having the bottom contact directly underneath the device instead of laterally to the side. This offers some potential advantages because it prevents current crowding and solves device area utilization problems. The bottom contact can help better dissipate heat from the device in the vertical structure because it is situated to conduct heat out of the active device area.

Other variations of group III-Nitride light emitters are micro LEDs and multi-wavelength light emitters. The reader can find relevant details in References [68–103]. The basic concepts about polarization, band-bending, etc. still apply to these device variations.

References

1 Zhou, S. et al. (2019). High quality GaN buffer layer by isoelectronic doping and its application to 365 nm InGaN/AlGaN ultraviolet light-emitting diodes. *App. Sur. Sci.* 471.

2 Kim, J. et al. (2013). Toward highly radiative white light emitting nanostructures: a new approach for dislocationeliminated GaN/InGaN core-shell nanostructures with a negligible polarization field. *Nanoscale* 1–3.

3 Bui, H. et al. (2019). Full-color InGaN/AlGaN nanowire micro light-emitting diodes grown by molecular beam epitaxy: a promising candidate for next generation micro displays. *Micromachines* 10 (8).

4 Nguyen, H.P.T., Zhang, S., Connie, A.T. et al. (2013). Breaking the carrier injection bottleneck of phosphor-free nanowire white light-emitting diodes. *Nano. Lett.* 13 (11).

5 Tchernycheva, M., Neplokh, V., Zhang, H. et al. (2015). Core-shell InGaN/GaN nanowire light emitting diodes analyzed by electron beam induced current microscopy and cathodoluminescence mapping. *Nanoscale* 7.

6 Tendille, F., De Mierry, P., Venn´egu`es, P. et al. (2014). Defect reduction method in (11–22) semipolar GaN grown on patterned sapphire substrate by MOCVD: toward heteroepitaxial semipolar GaN free of basal stacking faults. *J. Cryst. Growth* 404.

7 Ding, K., Avrutin, V., Özgür, Ü., and Morkoç, H. (2017). Status of growth of group III-Nitride heterostructures for deep ultraviolet light-emitting diodes. *Crystals* 7 (10).

8 Zhao, C., Tang, C.W., Cheng, G. et al. (2020). InGaN quantum dots with short exciton lifetimes grown on polar c-plane by metal-organic chemical vapor deposition. Mater. *Res. Express* 7 (11).

9 Pandey, A. et al. (2022). N-polar InGaN/GaN nanowires: overcoming the efficiency cliff of red-emitting micro-LEDs. *Photonics Res.* 10 (4).

10 Nguyen, H.P.T., Djavid, M., and Mi, Z. (2013). Nonradiative recombination mechanism in phosphor-free gan-based nanowire white light emitting diodes and the e_ect of ammonium sulfide surface passivation. *ECS Trans.* 53 (2).

11 Mehari, S., Cohen, D.A., Becerra, D.L. et al. (2018). Demonstration of enhanced continuous-wave operation of blue laser diodes on a semipolar 2021 GaN substrate using indium-tin-oxide/thin-p- GaN cladding layers. *Opt. Express* 26 (2).

12 Schweizer, H., Gräbeldinger, H., Dumitru, V. et al. (2002). Laterally coupled InGaN/GaN DFB laser diodes. *Phys. Status Solidi A* 192 (2).

13 Prasad, S., Mondal, R.K., Chatterjee, V., and Pal, S. (2019). Double-side step-graded AlGaN electron blocking layer for nearly droop-free GaNbased blue LEDs. *Superlattices Microstruct.* 132.

14 Philip, M.R., Bui, T.H.Q., Djavid, D.D.C.M. et al. (2018). Molecular beam epitaxial growth and device characterization of algan uv-b nanowire light-emitting diodes. *J. Adv. Opt. Photonics* 1 (1).

15 Banerjee, A., Frost, T., Jahangir, S. et al. (2013). InGaN/GaN self-organized quantum dot lasers grown by molecular beam epitaxy. *J. Cryst. Growth* 378.

16 Iida, D. and Ohkawa, K. (2021). Recent progress in red light-emitting diodes by III-nitride materials. *Semicond. Sci. Technol.* 37 (1).

17 Huang, Y. et al. (2021). Improved performance of gan-based ultraviolet leds with electron blocking layers composed of double-peak p-Type AlxGa1−xN/GaN superlattice layers. *IEEE Access* 9.

18 Mi, Z., Nguyen, H., Djavid, M. et al. (2014). High power phosphor-free InGaNGaNAlGaN core-shell nanowire white light emitting diodes on Si substrates. *ECS Trans.* 61 (5).

19 Dussaigne, A., Barbier, F., Damilano, B. et al. (2020). Full InGaN Red Light Emitting Diodes. *J. Appl. Phys.* 128 (13).

20 Kang, C.-M., Kong, D.-J., Shim, J.-P. et al. (2017). Fabrication of a vertically-stacked passive-matrix micro-LED array structure for a dual color display. *Opt. Express* 25 (3).

21 Yin, L., Bai, Y., Zhou, J. et al. (2015). The thermal stability performances of the color rendering index of white light emitting diodes with the red quantum dots encapsulation. *Opt. Mater.* 42.

22 Zhang, H., Cohen, D.A., Chan, P. et al. (2019). Continuous-wave operation of a semipolar InGaN distributed-feedback blue laser diode with a first-order indium tin oxide surface grating. *Opt. Lett.* 44 (12).

23 Nguyen, H.P.T., Djavid, M., Woo, S.Y. et al. (2015). Engineering the carrier dynamics of InGaN nanowire white light-emitting diodes by distributed p-AlGaN electron blocking layers. *Sci. Rep.* 5: 7744.

24 Asad, M., Wang, R., Ra, Y.-H. et al. (2019). Optically invariant InGaN nanowire light-emitting diodes on flexible substrates under mechanical manipulation. *Npj Flex. Electron.* 3: 16.

25 Wang, L., Yang, D., Hao, Z.B., and Luo, Y. (2015). Metal-organic-vapor phase epitaxy of InGaN quantum dots and their applications in light-emitting diodes. *Chin. Phys. B* 24 (6): 067303.

26 Song, J., Choi, J., Zhang, C., and Deng, Z. Xie Y and Han J. (2019). Elimination of stacking faults in semipolar GaN and light-emitting diodes grown on sapphire. *ACS Appl. Mater. Interfaces* 11 (36).

27 Philip, M.R., Choudhary, D.D., Djavid, M. et al. (2017). High efficiency green/yellow and red InGaN/AlGaN nanowire light-emitting diodes grown by molecular beam epitaxy. *J. Sci. Adv. Mater. Devices* 2 (2).

28 Weng, G. et al. (2016). Low threshold continuous-wave lasing of yellow-green InGaN-QD vertical cavity surface-emitting lasers. *Opt. Express* 24 (14).

29 Park, J.H., Kim, D.Y., Hwang, S. et al. (2013). Enhanced overall efficiency of GaInN-based light-emitting diodes with reduced efficiency droop by Al composition-graded AlGaN/GaN superlattice electron blocking layer. *Appl. Phys. Lett.* 103 (6).

30 Verma, S., Pandey, S.K., Pandey, S.K., and Mukherjee, S. (2015). Theoretical simulation of Hybrid II-O/III-N green light-emitting diode with MgZnO/InGaN/MgZnO heterojunction. *Mater. Sci. Semicond. Process* 31.

31 Slight, T.J., Yadav, A., Odedina, O. et al. (2017). InGaN/GaN laser diodes with high order notched gratings. *IEEE Photonics Technol. Lett.* 29 (23).

32 Zhao, H. et al. (2011). Approaches for high internal quantum efficiency green InGaN light-emitting diodes with large overlap quantum wells. *Opt. Express* 19 (S4).

33 Yu, H. et al. (2018). Progress and prospects of GaN-based VCSEL from near-UV to green emission. *Prog. Quantum. Electron* 57.

34 Khoury, M., Li, H., Kuritzky, L.Y. et al. (2017). 444 nm InGaN light emitting diodes on low-defect-density GaN templates on patterned sapphire. *Appl. Phys. Express* 10 (10).

35 Kuritzky, L.Y., Becerra, D.L., Abbas, A.S. et al. (2016). Chemically assisted ion beam etching of laser diode facets on nonpolar and semipolar orientations of GaN. *Semicond. Sci. Technol.* 31 (7).

36 Wang, R., Liu, X., Shih, I., and Mi, Z. (2015). High efficiency, full-color AlInGaN quaternary nanowire light emitting diodes with spontaneous core-shell structures on Si. *Appl. Phys. Lett.* 106: 261104.

37 Zhou, X., Tian, P., Sher, C.-W. et al. (2020). Growth, transfer printing and colour conversion techniques towards full-colour micro-LED display. *Prog. Quantum Electron* 71.

38 Shi, J.-W. et al. (2016). III-nitride based cyan light-emitting diodes with GHz bandwidth for high-speed visible light communication. *IEEE Electron Device Lett.* 37 (7).

39 Liu, X. et al. (2021). High efficiency InGaN nanowire tunnel junction green micro-LEDs. *Appl. Phys. Lett.* 119 (14).

40 Arteev, D.S. et al. (2021). Temperature-dependent luminescent properties of dual-wavelength InGaN LEDs. *J. Lumin.* 234.

41 Glaab, J., Ploch, C., Kelz, R. et al. (2015). Degradation of (InAlGa)Nbased UV-B light emitting diodes stressed by current and temperature. *J. Appl. Phys.* 118.

42 Ding, K., Avrutin, V., Izyumskaya, N. et al. (2019). micro-LEDs, a manufacturability perspective. *Appl. Sci.* 9 (6).

43 Xhao, C. et al. (2020). Low-efficiency droop InGaN quantum dot light-emitting diodes operating in the "green gap". *Photonics Res.* 8 (5).

44 Wang, L. et al. (2021). Green InGaN quantum dots breaking through efficiency and bandwidth bottlenecks of micro-LEDs. *Laser Photonics Rev.* 15: 200406.

45 Sadi, T. et al. (2014). On the correlation of the Auger generated hot electron emission and efficiency droop in III-N light-emitting diodes. *Appl. Phys. Lett.* 105.

46 Jiang, Y. et al. (2015). Realization of high-luminous efficiency InGaN light-emitting diodes in the "green gap" range. *Sci. Rep.* 5.

47 Zhao, H. and Tansu, N. (2010). Optical gain characteristics of staggered InGaN quantum well lasers. *J. Appl. Phys.* 107.

48 Rashidi, A. et al. (2017). High speed nonpolar InGaN/GaN LEDs for visible light communication. *IEEE Photon. Technol. Lett.* 29 (4).

49 Li, H. et al. (2020). Development of efficient semipolar InGaN long wavelength light-emitting diodes and blue laser diodes grown on a high quality semipolar GaN/sapphire template. *J. Phys. Photonics* 2.

50 Zhao, Y. et al. (2020). Enhanced P-Type GaN Conductivity by Mg Delta Doped AlGaN/GaN Superlattice Structure. *MDPI Mater.* 14 (1).

51 Hou, Y. (2019). Size effect of nanocavity on the performance of InGaN/GaN nanorod based plasmonic nanolaser. *J. Lumin.* 208.

52 Ferreira, R. et al. (2016). "High bandwidth GaN-based micro-LEDs for multi- Gbps visible light communications,". *IEEE Photonics Technol. Lett.* 28 (19).

53 Feezell, D.F., Speck, J.S., DenBaars, S.P., and Nakamura, S. (2013 Feb). Semi- polar InGaN/GaN light-emitting diodes for high-efficiency solid-state lighting. *J. Disp. Technol.* 9 (4): 190–198.

54 Goharizzi, A. and Atahyarizzadeh, G. (2022). The investigation of carrier mobility effect on the performance characteristics of the InGaN-Based Vertical Cavity Surface Emitting Laser (VCSEL) by solving the rate equations. *Micro Nanosyst.* 14 (4).

55 Slight, T.J., Odedina, O., Meredith, W. et al. (2016). InGaN/GaN distributed feedback laser diodes with deeply etched sidewall gratings. *IEEE Photonics Technol. Lett.* 28 (24).

56 Jiang, H. and Lin, J. (2013). Nitride micro-LEDs and beyond-a decade progress review. *Opt. Express* 21.

57 Mei, Y. et al. (2021). Low-threshold wavelength-tunable ultraviolet vertical-cavity surface-emitting lasers from 376 to 409 nm. *Fundam. Res.* 1 (6).

58 Slight, T.J., Stanczyk, S., Watson, S. et al. (2018). Continuous-wave operation of (Al, In) GaN distributed-feedback laser diodes with high-order. *Appl. Phys. Express* 11 (11).

59 Wu, Y. et al. (2020). Full-Color Realization of Micro-LED Displays. *Nanomaterials* 10 (12).

60 Hofmann, R., Gauggel, H.P., Griesinger, U.A. et al. (1996). Realization of optically pumped second-order GaInN-distributed-feedback lasers. *Appl. Phys. Lett.* 69.

61 Frost, T., Su, G.L., Hazari, A. et al. (2017). Red and near-Infrared III-Nitride quantum dot lasers. *IEEE J. Sel.Top. Quantum Electron* 23 (6).

62 Kishino, K., Nagashima, K., and Yamano, K. (2012). Monolithic integration of InGaN-based nanocolumn light-emitting diodes with different emission colors. *Appl. Phys. Express* 6: 012101.

63 Nguyen, H.P.T., Cui, K., Zhang, S. et al. (2012). Controlling electron overflow in phosphor-free InGaN/GaN nanowire white light-emitting diodes. *Nano. Lett.* 12 (3).

64 Wang, Z., Shan, X., Cui, X., and Tian, P. (2020). Characteristics and techniques of GaN-based micro-LEDs for application in next-generation display. *J. Semicond.* 41 (4).

65 Holguín-Lerma, J.A., Ng, T.K., and Ooi, B.S. (2019). Narrow-line InGaN/GaN green laser diode with high-order distributed-feedback surface grating. *Appl. Phys. Express* 12 (4).

66 Saleem, M.F. et al. (2021). Factors affecting surface plasmon coupling of quantum wells in nitride-based LEDs: a review of the recent advances. *Nanomaterials* 11 (5).

67 Xu, H. et al. (2020). Green VCSELs based on nitride semiconductors. *Jpn. J. Appl. Phys.* 59.

68 Cho, I.W., Lee, B., Ryu, M.Y. et al. (2021). Luminescence properties of InGaN/GaN light-emitting diodes with violet, blue, and green emission. *J. Korean Phys. Soc.* 78.

69 Zhao, Y., Fu, H., Wang, G.T., and Nakamura, S. (2018). Toward ultimate efficiency: progress and prospects on planar and 3D nanostructured nonpolar and semipolar InGaN light-emitting diodes. *Adv. Opt. Photon* 10 (1).

70 Mehari, S., Cohen, D.A., Becerra, D.L. et al. (2019). Semipolar InGaN blue laser diodes with a low optical loss and a high material gain obtained by suppression of carrier accumulation in the pwaveguide region. *Jpn. J. Appl. Phys.* 58 (2).

71 Jiang, B. et al. (2020). The optimization of surface plasmon coupling efficiency in InGaN/GaN nanowire based nanolasers. *Applied Phys. Express* 13 (8).

72 Leung, B., Wang, D., Kuo, Y.-S. et al. (2014). Semipolar (201) GaN and InGaN quantum wells on sapphire substrates. *Appl. Phys. Lett.* 104.

73 Wei, X., Al Muyeed, S.A., Xue, H. et al. (2022). Near-infrared electroluminescence of algan capped ingan quantum dots formed by controlled growth on photoelectrochemical etched quantum dot templates. *Photonics Res.* 10 (1).

74 Liu, Y. et al. (2020). High-brightness InGaN/GaN Micro-LEDs with secondary peak effect for displays. *IEEE EDL* 41 (9).

75 Terao, K. et al. (2021). Blue and green GaN-based vertical-cavity surface-emitting lasers with AlInN/GaN DBR. *Proceedings of the SPIE* 11686.

76 Zhao, H., Liu, G., and Tansu, N. (2010). Analysis of InGaN-delta-InN quantum wells for light-emitting diodes. *Appl. Phys. Lett.* 97.

77 Nguyen, H.P.T., Cui, K., Zhang, S. et al. (2011). Fullcolor InGaN/GaN dot-in-a-wire light emitting diodes on silicon. *Nanotechnology* 22: 445202.

78 Myzaferi, A., Reading, A.H., Cohen, D.A. et al. (2016). Transparent conducting oxide clad limited area epitaxy semipolar III-nitride laser diodes. *Appl. Phys. Lett.* 109 (6).

79 Masui, S., Tsukayama, K., Yanamoto, T. et al. (2006). CW operation of the first-order AlInGaN 405 nm distributed feedback laser diodes. *Jpn. J. Appl. Phys.* 45.

80 Zhao, C., Tang, C.W., Lai, B. et al. (2020). Low-efficiency-droop ingan quantum dot light emitting diodes operating in the "green gap". *Photonics Res.* 8 (5).

81 Li, Y. et al. (2018). High-performance vertical GaN-based near-ultraviolet light-emitting diodes on Si substrates. *J. Mater. Chem.* 42.

82 Tang, Y. et al. (2022). GaN-based distributed feedback laser diodes grown on Si. *J. Phys. D* 55 (19).

83 Zhang, X. et al. (2022). Explorations on growth of blue-green-yellow-red InGaN quantum dots by plasma-assisted molecular beam epitaxy. *MDPI Nanomater.* 12 (5).

84 Moseley, M.W., Allerman, A.A., Crawford, M.H. et al. (2015). Detection and modeling of leakage current in AlGaN-based deep ultraviolet light-emitting diodes. *J. Appl. Phys.* 117 (9).

85 Holguin-Lerma, J. et al. (2020). 480-nm distributed-feedback InGaN laser diode for 10.5-Gbit/s visible-light communication. *Opt. Lett.* 45 (3).

86 Lee, V.W., Twu, N., and Kymissis, I. (2016). micro-LED technologies and applications. *Inf. Disp.* 32 (6).

87 Kishino, K., Yanagihara, A., Ikeda, K., and Yamano, K. (2015). Monolithic integration of four-colour InGaN-based nanocolumn LEDs. *Electron. Lett.* 51: 852–854.

88 Zhang, C. et al. (2019). Distributed bragg reflector for GaN-based vertical cavity surface emitting lasers. *Appl. Sci.* 9 (1593).

89 Schulz, S. and O'Reilly, E.P. (2010). Theory of reduced built-in polarization field in nitride-based quantum dots. *Phys. Rev. B* 82.

90 Deng, Z. et al. (2019). InGaN/GaN distributed feedback laser diodes withsurface gratings and sidewall gratings. *Micromachines* 10 (10).

91 Fujioka, A., Asada, K., Yamada, H. et al. (2014). High-output-power 255/280/310 nm deep ultraviolet light-emitting diodes and their lifetime characteristics. *Semicond. Sci. Technol.* 29 (8).

92 Ren, Z., Yu, H., Liu, Z. et al. (2019). Band engineering of III-nitride-based deep-ultraviolet light-emitting diodes: a review. *J. Phys. D, Appl. Phys.* 53 (7).

93 Hahn, C. et al. (2011). Epitaxial growth of ingan nanowire arrays for light emitting diodes. *ACS Nano*. 5 (5): 3970–3976.

94 Shen, C. et al. (2018). Semipolar InGaN quantum-well laser diode with integrated amplifier for visible light communications. *Opt. Express* 26 (6).

95 Lu, L., Ding, G.G., Zhang, Y. et al. (2018). Improved performance of AlGaN-based deep ultraviolet light-emitting diode using modulated-taper design for p-AlGaN layer. *Semicond. Sci. Technol.* 33 (3).

96 Xie, W. et al. (2019). Narrow linewidth distributed bragg reflector based InGaN/GaN laser. *Micromachines* 10.

97 Huang, Y., Hsiang, E.-L., Deng, M.-Y., and Wu, S.-T. (2020). Mini-LED, micro-LED and OLED displays: present status and future perspectives. *Light Sci. Appl.*

98 Liu, L., Wang, L., Liu, N. et al. (2012). Investigation of the light emission properties and carrier dynamics in dual-wavelength InGaN/GaN multiple-quantumwell light emitting diodes. *J. Appl. Phys.* 9.

99 Chang, Y.-A., Kuo, Y.-T., Chang, J.-Y., and Kuo, Y.-K. (2012). Investigation of InGaN green light-emitting diodes with chirped multiple quantum well structures. *Opt. Lett.* 37 (12).

100 Ryu, H.-Y., Choi, I.-G., Choi, H.-S., and Shim, J.-I. (2013). Investigation of light extraction efficiency in AlGaN deep-ultraviolet light-emitting diodes. *Appl. Phys. Exp.* 6 (6).

101 Islam, S.M., Protasenko, V., Rouvimov, S. et al. (2016). Sub-230 nm deep-UV emission from GaN quantum disks in AlN grown by a modified Stranski–Krastanov mode. *Jpn. J. Appl. Phys.* 55 (55).

102 Jiang, X., Zheng, C., Mo, C. et al. (2019). Study on the performance of InGaN-based green led by designing different preparing layers. *Opt. Mater.* 89.

103 Lv, W., Wang, L., Wang, J. et al. (2013). Green and red light-emitting diodes based on multilayer InGaN/GaN n dots grown by growth interruption method. *Jpn. J. Appl. Phys.* 52.

104 Y.Mei et al. (2017). Tunable InGaN quantum dot microcavity light emitters with 129nm tuning range from yellow-green to violet. *Appl. Phys. Lett.* 111.

6

Inter Sub-band Devices

Quantum Cascade Lasers

While optical lasers (based on quantum wells, quantum wires, and quantum dots [QDs]) are of the greatest interest in the group III-Nitride material system, infrared emitters can also be fabricated in this material system. We will discuss these infrared emitters very briefly.

The quantum cascade laser (QCL) relies on inter sub-band transitions between one quantum well and an adjacent coupled quantum well. A fast tunneling time from the biasing lead to a sub-band level is necessary to define the upper state of the lasing transition. A relatively slower transition from this sub-band and a sub-band with wave function peaking in the second well layer defines the lasing transition. A fast transition from this second sub-band level to a third sub-band level ensures that a population inversion is maintained between the first sub-band (centered primarily in the leftmost quantum well) and the second sub-band (centered on the second quantum well from the left), as shown in Figure 6.1.

The influence of scattering mechanisms, especially interface roughness scattering, on QCL performance shows [1] that there is considerable scope for gain improvement by improving interface quality in the quantum-well regions of the QCL. Figure 6.2 shows typical AlGaN QCL structure and band diagrams. References [2–7] contain details about typical group III-Nitride inter sub-band emitters.

Superlattice Avalanche Transit-time Diode Terahertz Source

Avalanche transit-time diodes with mixed tunneling and avalanche transport can be used to generate terahertz oscillators [8]. These mixed-impact ionization tunneling avalanche transit-time (MITATT) diodes need careful conductance optimization in the top and bottom layers to get good device efficiencies. Figure 6.3 shows an MITATT superlattice diode from the literature.

Infrared Photodetectors

Indium gallium nitride (InGaN) infrared photodetectors can in principle be fabricated using doped InGaN quantum-well structures. The indium (In) composition in the InGaN quantum-well layer, the composition of the barrier layer and the thickness of the quantum-well layer need to be chosen to give inter sub-band absorption in the range for which the infrared detector is designed. After the

Group III-Nitride Semiconductor Optoelectronics, First Edition. C. Jayant Praharaj.

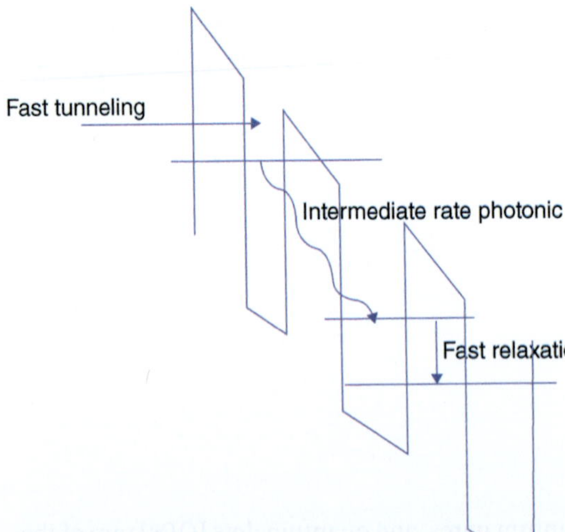

Figure 6.1 A quantum cascade laser (QCL) structure based on a nitrogen (N)-face indium gallium nitride (InGaN) multi quantum-well design. For In-face or Ga-face growth, more careful optimization of compositions and widths are needed.

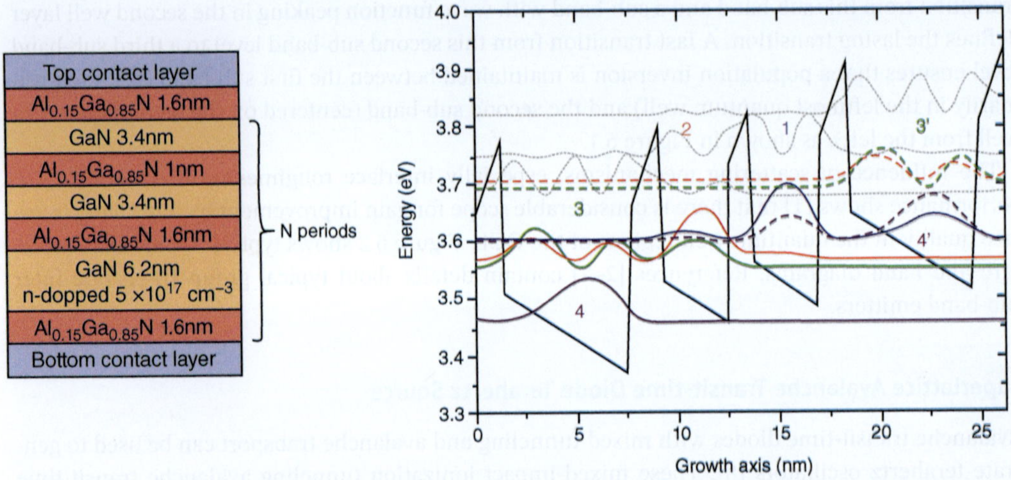

Figure 6.2 Structure and energy levels of an aluminum gallium nitride (AlGaN)/gallium nitride (GaN) QCL. From Reference [1], used with permission.

excited electron is created in one of the sub-bands, a fast tunneling transition is required to get a good photocurrent. This requires good barrier layer design. The barrier must be confined to give the required inter sub-band energy and be thin enough to give good tunneling under bias.

Aluminum gallium nitride (AlGaN)/gallium nitride (GaN) superlattices can be used to obtain photodetectors in the near-infrared range, as shown in Figures 6.4 and 6.5. However, the AlGaN barrier layer is difficult to dope and get good layer conductivity due to the large effective mass of electrons, which leads to large donor ionization energies. Degenerate doping with a Mott metal-insulator transition and the formation of dopant bands needs to be used and optimized to get good detector efficiencies and signal-to-noise ratios (SNRs).

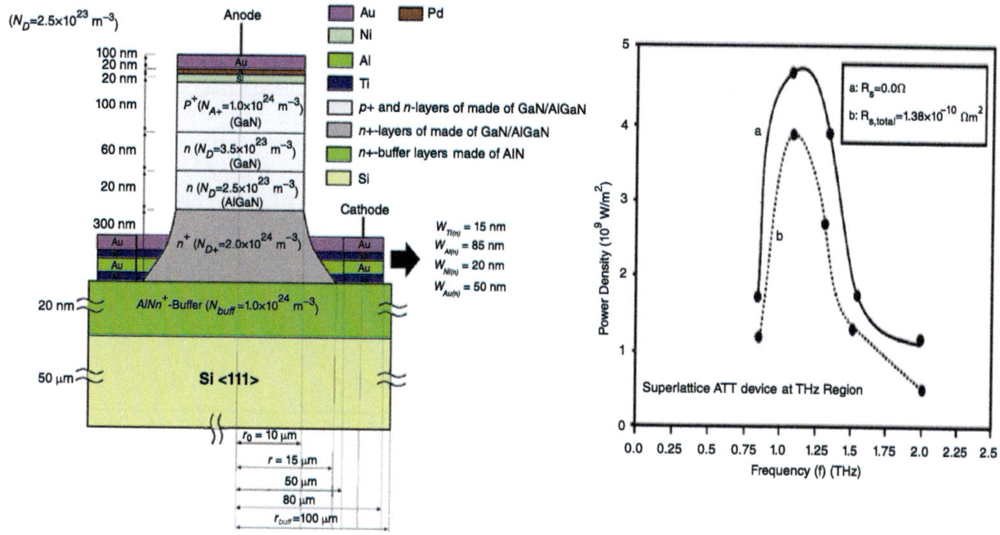

Figure 6.3 MITATT superlattice diode structure and power output. From Reference [9], used with permission.

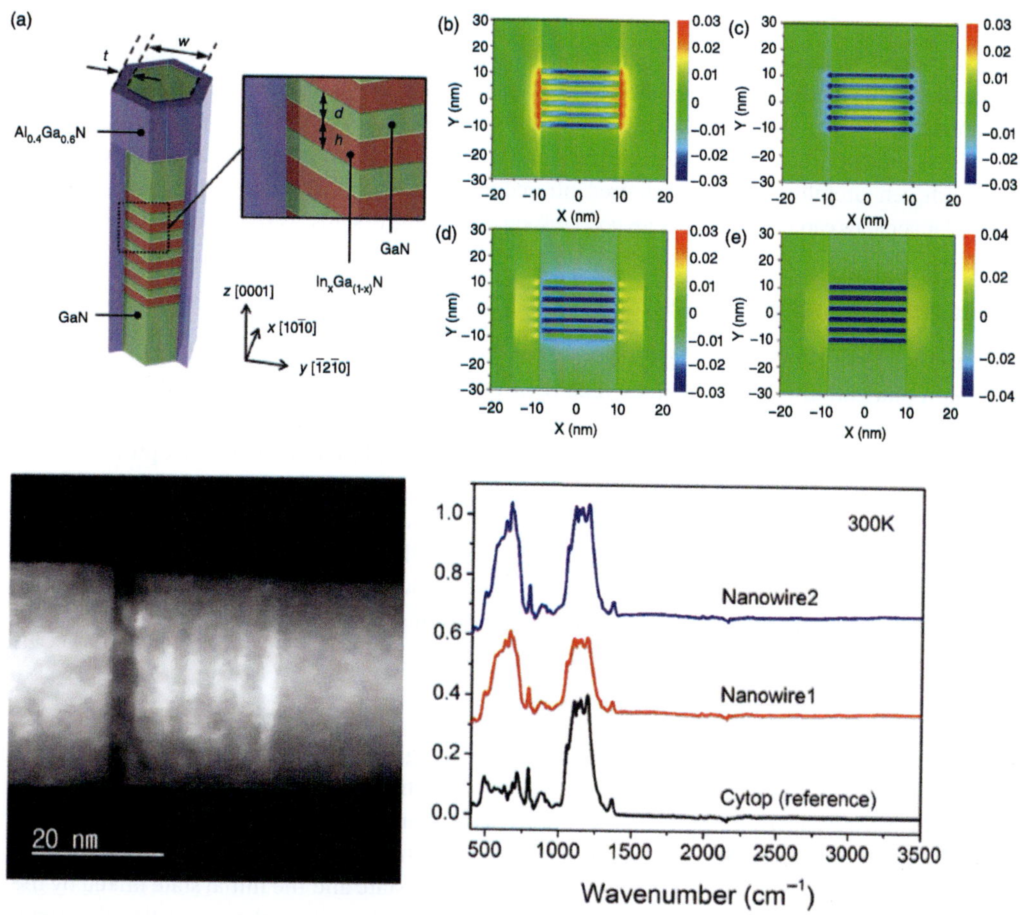

Figure 6.4 Nanodisks for infrared photon absorption. Structure and strain components (top). Transmission electron microscopy (TEM) cross-section and absorbance (bottom). From Reference [10], used with permission.

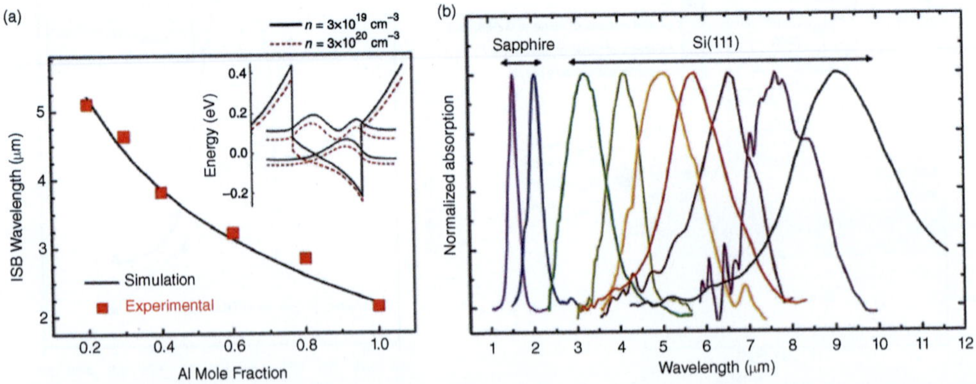

Figure 6.5 Variation of inter sub-band wavelength with aluminum (Al) content in AlGaN/GaN superlattice. From Reference [11], used with permission.

Superlattice Photodetectors

Superlattices from aluminum nitride (AlN) and GaN can be used to obtain bands separated by energies that correspond to infrared wavelengths. *c*-Axis superlattices, which are the most common, have spontaneous and piezoelectric polarization effects. There is polarization-induced sheet charge density at each interface between AlN and GaN. This gives built-in electric fields in each layer. This has a non-trivial effect on superlattice miniband dispersions and energies. Therefore, composition and layer widths must be chosen carefully to obtain the desired infrared photon absorption characteristics. Non-polar and semi-polar versions can be used to obtain good responses in some designs, as shown in Figure 6.6.

Quantum-well Inter Sub-band Photodetectors

Quantum-well sub-band energy levels are determined by effective masses, conduction band offsets, and polarization-induced charges. The SNR of the photodetectors is affected by the series resistance provided by the *n*-type or *p*-type doped collection layers. Therefore, the wavelength and energy range of InGaN alloys and AlGaN alloys need to be explored to obtain the optimal design for any application. The effective mass range for InGaN alloy spans 0.07–0.25, while that of AlGaN alloy spans 0.25–0.4. Therefore, there is quite a bit of flexibility in designing the wavelength range of inter sub-band transitions in quantum wells made from this material system.

The composition and dimensions of the multi quantum-well layers become the design parameters to optimize to meet the wavelength range specification and the efficiency metrics.

Multiple Quantum-well Inter Sub-band Photodetectors

Multiple quantum wells are intermediates between single quantum wells and superlattices. Multiple quantum wells have electron and hole wave functions that span several quantum wells. Different sub-bands have different concentrations of electron density in the various well regions of the multi quantum-well device. The quantum mechanical transition involved in absorption has a parity or symmetry requirement, which states that the final state and the initial state linked by the photon absorption must have opposite parity. To get the best absorption characteristics, the parity difference must be right between initial and final states, and in addition, there must be good overlap between initial and final states. This means careful optimization is needed to obtain the best sensitivity and SNR in multi quantum-well infrared detectors.

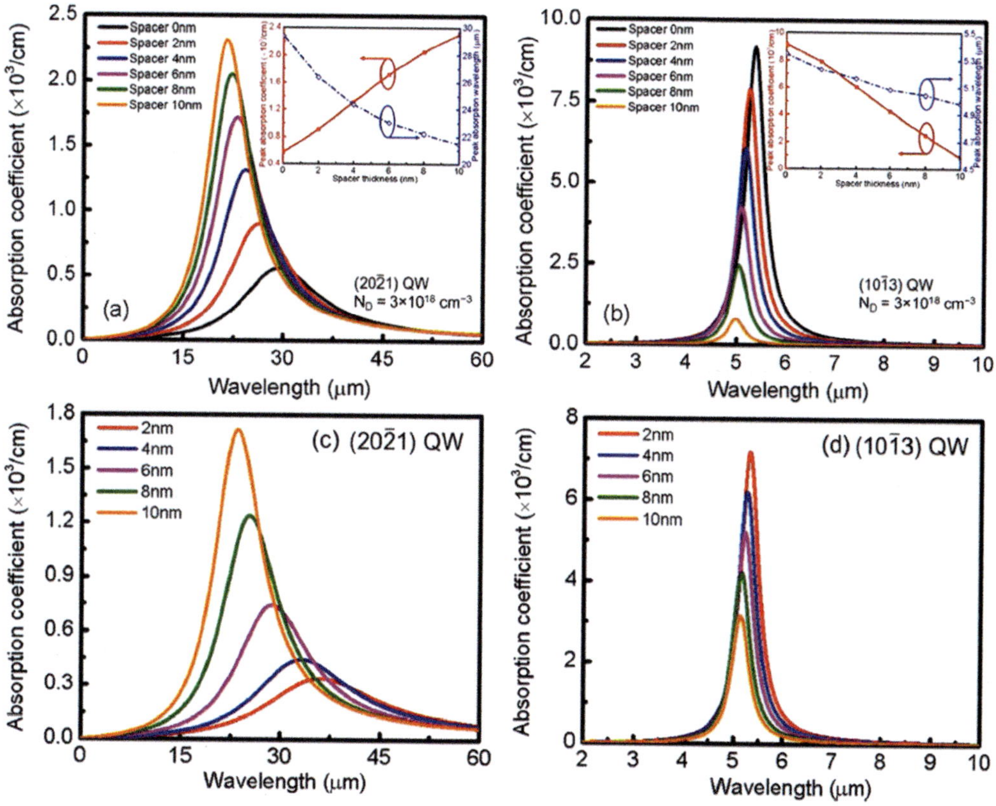

Figure 6.6 Inter sub-band absorption in non-polar and semi-polar AlGaN/GaN quantum well. From Reference [11], used with permission.

Broad Response Band Quantum Cascade Detector

Polarization-induced fields in AlGaN/GaN quantum cascade detectors lead to absorption characteristics different from inter sub-band detectors without polarization-induced fields. For example, transitions from the ground state to the first excited state and from the ground state to the second excited state are possible. Figure 6.7 shows the typical structure of a broadband quantum cascade detector and its responsivity as a function of wavelength. The response spans the range from mid-infrared to the visible.

Quantum Cascade Detectors on Different Substrates

Since the substrate decides a lot of factors, like the dislocation density in the device layers, detectors built on different substrates show different performance due to dislocation and strain effects. The cost aspect makes silicon attractive as a substrate for quantum cascade detectors. However, technological optimization requires growth method optimization to minimize dislocation densities.

Group III nitride on silicon (Si) photodetectors offer the scope for monolithic integration with Si electronics because epitaxial growth on Si substrate can be optimized for optimum device performance, although challenges still remain. Dislocations can degrade the performance of quantum cascade detectors because they cause Shockley–Read–Hall type trap-assisted recombinations, which can

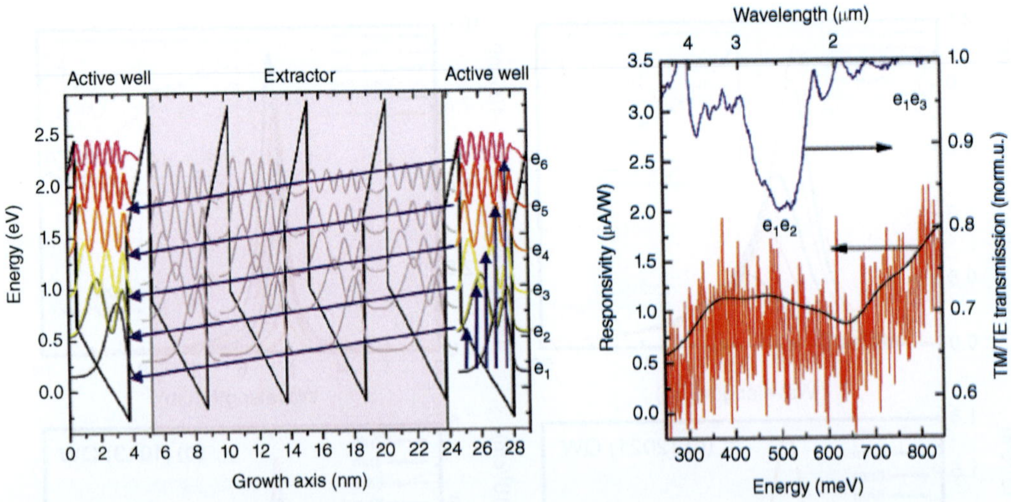

Figure 6.7 Broadband quantum cascade detector structure and responsivity. From Reference [12], used with permission.

degrade the current response. Dislocation-assisted recombination leads to lowered quantum efficiencies and lower SNRs of the overall circuit. The use of template buffer layers, like superlattice buffer layers, can help lower the dislocation density because template buffer layers can help absorb strain and increase the thickness limits for dislocation formation. The reader can find detailed reports on inter sub-band detectors and absorption physics in References [13–32].

References

1 Cheng, J. et al. (2019). Dominant influence of interface roughness scattering on the performance of GaN terahertz quantum cascade lasers. *Nano Res. Lett.* 14 (206).

2 Song, Y., Bhat, R., Huang, T.Y. et al. (2014). III-Nitride quantum cascade detector grown by metal organic chemical vapor deposition. *Appl. Phys. Lett.* 105 (18).

3 Kang, D.D., Inoue, T., Asano, T., and Noda, S. (2019). Electrical modulation of Narrowband GaN/AlGaN quantum-well photonic crystal thermal emitters in mid-wavelength infrared. *ACS Photonics* 6 (6).

4 Lim, C.B., Ajay, A., Bougerol, C. et al. (2016). Effect of doping on the far-infrared intersubband transitions in nonpolar m-plane GaN/AlGaN heterostructures. *Nanotechnology* 27 (14).

5 Burmistrov, E. et al. (2021). Piezoelectric relaxation of two-dimentional electron gas in InGaN/GaN heterostructures with quantum wells. *Springer Nat. Appl. Sci.* 3: 758.

6 Machhadani, H. et al. (2009). GaN/AIGaN intersubband optoelectronic devices. *N. J. Phys.* 11.

7 Piprek, J. et al. (2022). GaN-based bipolar cascade lasers with 25 nm wide quantum wells. *Opt. Quantum Electron* 54.

8 Sudradjat, F., Zhang, W., Woodward, J. et al. (2012). Far-infrared intersubband photodetectors based on double-step Ill-nitride quantum wells. *Appl. Phys. Lett.* 100.

9 Adhikari, S. et al. (2020). Super-lattice GaN/AlxGa1-xN nanoscale MITATT oscillator as terahertz radiation source: novel application in breast cancer imaging. *Sensors Int.* 1.
10 Xing, Z. et al. (2022). Mid-infrared photon sensing using InGaN/GaN nanodisks via intersubband absorption. *Nat. Sci. Rep.* 22: 4301.
11 Mudiyanselage, M. (2022). Intersubband transitions in nonpolar and semipolar III-nitrides: materials, devices, and applications. *J. Appl. Phys.* 131 (21).
12 Quach, P. et al. (2020). A GaN/AlN quantum cascade detector with a broad response from the mid-infrared (4.1 μm) to the visible (550 nm) spectral range. *Appl. Phys. Lett.* 116 (17).
13 Kandaswamy, P.K. et al. (2010). Effect of doping on the mid-infrared intersubband absorption in GaN/AlGaN superlattices grown on Si(111) templates. *Appl. Phys. Lett.* 96.
14 Kandaswamy, P.K. et al. (2008). GaN/AlN short-period superlattices for intersubband optoelectronics: a systematic study of their epitaxial growth, design, and performance. *J. Appl. Phys.* 104.
15 Beeler, M. et al. (2014). Intraband absorption in self-assembled Ge-doped GaN/AlN nanowire heterostructures. *Nano Lett.* 14 (3).
16 Sudradjat, F.F., Zhang, W., Woodward, J. et al. (2012). Far-infrared intersubband photodetectors based on double-Step III-Nitride quantum wells. *Appl. Phys. Lett.* 100.
17 Lim, C.B., Beeler, M., Ajay, A. et al. (2016). Short-wavelength, mid- and farinfrared intersubband absorption in nonpolar GaN/Al(Ga)N heterostructures, Jpn. *J. Appl. Phys.* 55.
18 Quach, P., Liu, S.F., Jollivet, A. et al. (2020). A GaN/AlN quantum cascade detector with a broad response from the Mid-Infrared (4.1 _m) to the Visible (550 nm) Spectral range. *Appl. Phys. Lett.* 116.
19 Cheng, J., Quach, P., Wang, D. et al. (2019). Dominant influence of interface roughness scattering on the performance of GaN terahertz quantum cascade lasers. *Nanoscale Res. Lett.* 14.
20 Giannoccaro, G., de Leonardis, F., and Passaro, V.M.N. (2016). Modeling of the interminiband absorption coefficient in InGaN quantum dot superlattices. *Photonics* 3 (1).
21 Song, Y., Bhat, R., Huang, T.Y. et al. (2014). III-Nitride quantum cascade detector grown by metal organic chemical vapor deposition. *Appl. Phys. Lett.* 105.
22 Sakr, S., Giraud, E., Dussaigne, A. et al. (2012). Two-Color GaN/AlGaN quantum cascade detector at short infraredwavelengths of 1 and 1.7 _m. *Appl. Phys. Lett.* 100.
23 Chuang, S.L. (2009). *Physics of Optoelectronic Devices.* John Wiley and Sons.
24 Durmaz, H., Nothern, D., Brummer, G. et al. (2016). Terahertz intersubband photodetectors Based on Semipolar III-Nitride quantum wells. *Appl. Phys. Lett.* 108.
25 Machhadani, H., Beeler, M., Sakr, S. et al. (2013). Systematic study of near-infrared intersubband absorption of polar and semipolar GaN/AlN quantum wells. *J. Appl. Phys.* 113.
26 Chen, G. et al. (2015). Intersubband transition in GaN/InGaN multiple quantum wells. *Sci. Rep.* 5.
27 Pesach, A., Gross, E., Huang, C.-Y. et al. (2013). Non-polar m-plane intersubband based InGaN/(Al)GaN quantum well infrared photodetectors. *Appl. Phys. Lett.* 103 (2).
28 Kandaswamy, P.K. et al. (2009). Midinfrared intersubband absorption in GaN/AlGaN superlattices on Si(111) templates. *Appl. Phys. Lett.* 95 (14).
29 Guillot, F. et al. (2006). Si-doped GaN/AlN quantum dot superlattices for optoelectronics at telecommunication wavelengths. *J. Appl. Phys.* 100 (4).
30 Nguyen, T., Dzuba, B., Cao, Y. et al. (2021). Mid-infrared intersubband absorption in strain-balanced nonpolar (In)AlGaN/InGaN multi-quantum wells. *Opt. Mater. Express* 11 (9).
31 Beeler, M., Trichas, E., and Monroy, E. (2013). III-nitride semiconductors for intersubband optoelectronics: a review. *Semicond. Sci. Technol.* 28 (7).
32 Chen, G. et al. (2013). Effect of polarization on intersubband transition in AlGaN/GaN multiple quantum wells. *Appl. Phys. Lett.* 102 (19).

7

Photodetectors

Ultraviolet Photodetectors

Since the bandgap of gallium nitride (GaN; 3.43 eV) and that of aluminum nitride (AlN; 6.2 eV) are in the ultraviolet (UV), this material system offers the possibility of fabricating detectors that are sensitive to UV light without being sensitive to visible light. Further, the 0.7 eV bandgap of indium nitride (InN) means that the indium gallium nitride (InGaN) alloy can be used to obtain photodetectors in the visible range as well.

All the group III-Nitride semiconductors have direct bandgaps. This means that absorption coefficients can be high enough to obtain good absorption with lower device widths. The bandgaps of aluminum gallium nitride (AlGaN) and InGaN alloy semiconductors are also of the direct bandgap variety according to most studies. Therefore, alloying can be used to obtain the desired spectral response.

Complex Dielectric Function

The absorption properties of InGaN and AlGaN alloys are determined by the imaginary part of the frequency-dependent dielectric function. Figures 7.1 and 7.2 show typical dielectric functions in group III nitrides.

The absorption coefficient is given in terms of the imaginary part of the dielectric constant by

$$\alpha = \frac{\omega}{n_r c} \frac{\varepsilon_2}{\varepsilon_0} \tag{7.1}$$

The absorption coefficient is ideally zero below the bandgap and rises starting from the bandgap energy. There is an excitonic peak below the bandgap edge. For high indium percentage InGaN, the exciton binding energy is lower than the thermal energy at a room temperature of 25.6 meV. However, for AlGaN, the exciton binding energy is larger and can play an important role in some devices. The calculated and experimentally determined dielectric functions can be fit to analytical expressions for use in device simulation. The actual dielectric functions and absorption coefficients are anisotropic for wurtzite InGaN and AlGaN, and this anisotropy can affect device operation depending on the crystal orientation used to build device layers.

Group III-Nitride Semiconductor Optoelectronics, First Edition. C. Jayant Praharaj.

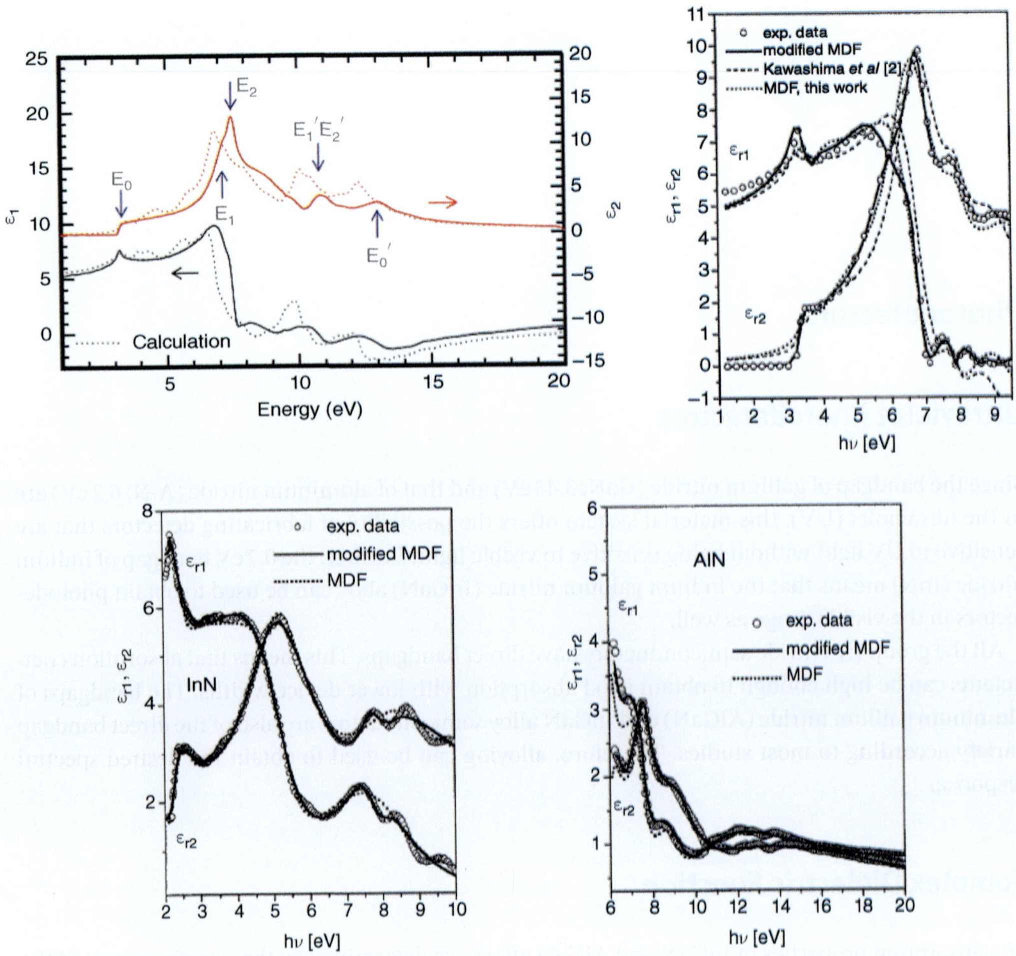

Figure 7.1 Real and imaginary parts of the dielectric function versus frequency for cubic GaN, hexagonal (wurtzite) GaN, hexagonal (wurtzite) InN, and hexagonal(wurtzite) AlN – from References [1 and 2], used with permission.

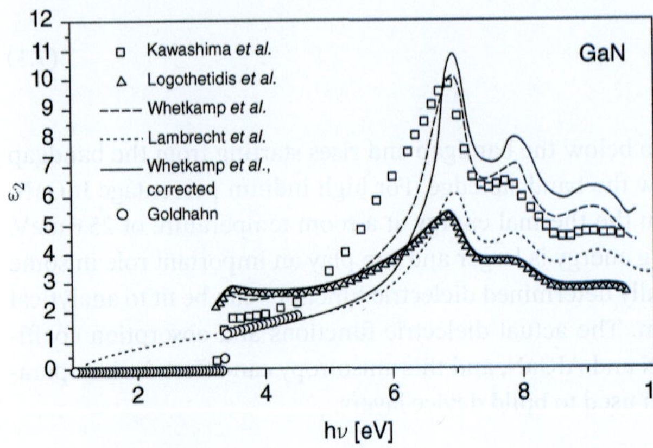

Figure 7.2 Theoretical and experimental dielectric function for hexagonal (wurtzite) GaN – from Reference [3], used with permission.

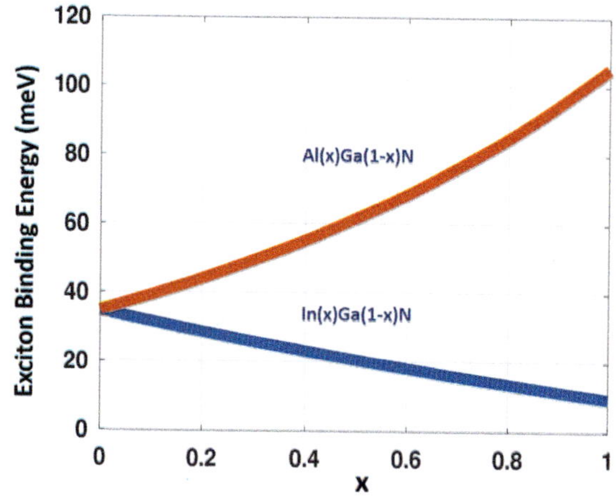

Figure 7.3 Exciton binding energies versus composition for InGaN and AlGaN.

The imaginary part of the dielectric function and the frequency-dependent absorption coefficient have different behaviors just above the bandgap and far above the bandgap. Near the bandgap, the absorption coefficient tends to depend on the square root of photon energy minus the bandgap energy. Far above the bandgap, the absorption coefficient shows a more complex structure because the imaginary part of the dielectric constant depends on the conduction band as well as the valence-band density of states, and these have complex structures far above the bandgap.

Photodetectors can be designed to detect optical radiation or incident photon radiation that is just above the bandgap or well above the bandgap. When the photon energy is well above the bandgap, electrons can be generated well above the conduction-band minimum and holes can be generated well below the valence-band maximum. When the generated electrons are well above the band minimum, they can relax their energy by emitting longitudinal optical (LO) phonons and move closer to the band minimum. In photodetectors there is a built-in electric field that makes the photogenerated electrons drift and get collected as current. If the transit time of the electrons is smaller than the LO phonon-mediated energy relaxation time, the electrons will drift while remaining far above the band minimum and possibly see different effective masses, saturation velocities, and mobilities. Similar conclusions hold for photogenerated holes that are generated far below the valence-band maximum.

Photogenerated electrons and holes can also form excitonic bound states due to Coulomb attraction between electrons and holes. Figure 7.3 shows exciton binding energies in various group III-Nitride semiconductors. The reader must keep in mind that exciton binding may become relevant to transport and device performance in some device designs. In these cases, more complicated device transport equations must be used.

Basic Principle of Operation

The semiconductor photodetector usually works by getting a photogenerated current in a reverse biased *p-n* diode, a reverse biased *p-i-n*-type diode, a metal-semiconductor-metal structure, or a similar device. For a *p-n* diode photodetector, the charge continuity equation for holes is

$$\frac{\partial p_n}{\partial t} = G(x,t) - \frac{\delta p_n}{\tau_p} - \frac{1}{q}\frac{\partial}{\partial x} J_p(x) \tag{7.2}$$

where $p_n = p_{n0} + \delta p_n$ is the total hole concentration in the n-type region. The generation rate of electron-hole pairs can be written in terms of the absorption coefficient as

$$G(x,t) = \alpha(x)\frac{I(x,t)}{h\nu} \tag{7.3}$$

where $I(x,t)$ is the intensity, h is Planck's constant, and ν is frequency. Since the minority hole current density in the n-region is dominated by the diffusion current, it can be written as

$$J_p(x) = -qD_p\frac{\partial p_n}{\partial x} \tag{7.4}$$

In steady state, the current continuity equation becomes

$$D_p\frac{\partial^2}{\partial x^2}\delta p_n - \frac{\delta p_n}{\tau_p} = -G_0 \tag{7.5}$$

Using the commonly used approximation of a flat quasi-Fermi level across the depletion region (which gives a good agreement with experiments at moderate bias), we obtain, upon solving the differential equation above,

$$\delta p_n(x) = \left[p_{n0}\left(exp\left(\frac{qV}{kT}\right) - 1\right) - G_0\tau_p\right]e^{-\frac{x-x_n}{L_p}} + G_0\tau_p \tag{7.6}$$

A similar expression holds for the excess n-type electron concentration in the p-type quasi-neutral region. When the recombination-generation in the depletion region is neglected, this leads to the following expression for the total current across the p-n diode under illumination

$$I = A\left[J_n(-x_p) + J_p(x_n)\right] \tag{7.7}$$

$$I = I_0\left(e^{\frac{qV}{kT}} - 1\right) - qAG_0(L_p + L_n) \tag{7.8}$$

The usual I–V curve of the p-n diode without illumination is shifted downward due to illumination, and we obtain a short-circuit current at zero bias. In a real photodiode, the generation rate due to light is not uniform across all the diode regions. Instead, the generation rate decreases exponentially from the metal interface

$$G(x) = G_0\exp(-ax) \tag{7.9}$$

Numerical solutions typically become necessary when the generation rate is a function of position.

Heterojunction devices with GaN as one of the layers have given good photoresponses in the ultraviolet (UV), as shown in Figure 7.4. The second layer of the heterojunction is chosen in a manner such that the absorption characteristics of both layers are optimal for getting the best response from the incident radiation.

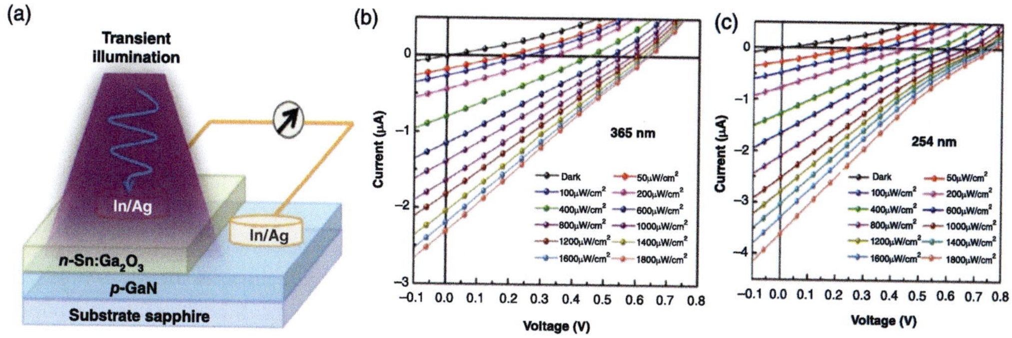

Figure 7.4 Photoresponse of a heterojunction photodetector with GaN as one of the layers. From Reference [4] – used with permission.

Metal–Semiconductor–Metal (MSM) Photodetector

A metal–semiconductor–metal (MSM) photodetector uses the drift of photogenerated electrons and holes to obtain a photocurrent, which can then be converted into a voltage by a signal processing circuit. The complex dielectric constant determines the generation rate of electron-hole pairs in the semiconductor. The absorption rate is proportional to the imaginary part of the complex dielectric function.

The photogenerated electrons and holes drift in opposite directions and lead to a photocurrent. A large enough electric field is typically maintained across the semiconductor so that the drift velocity is the LO phonon-limited saturation velocity of the electrons and holes in the semiconductor. Since the valence-band holes have a higher mass compared to the conduction-band electrons, the saturation drift velocity of the holes is smaller compared to the electrons. Also, there are three types of holes – i) the heavy holes, ii) the light holes, and iii) the split-off holes. The heavy holes have the highest mass and the lowest saturation drift velocity. Also, the effective masses of electrons and holes are anisotropic. The effective mass along the *c*-axis is different from the effective mass perpendicular to the *c*-axis. This also leads to different saturation velocities for different electric field directions compared to the crystal axes of symmetry. The LO phonon frequency increases from InN through GaN to AlN. The saturation velocity shows a weak upward trend in the same order.

The transit time of holes across the semiconductor layer determines the response speed of the MSM photodetector. The transit time is given by

$$\tau_{tr} = \frac{L}{v_{sat}} \tag{7.10}$$

The saturation velocity of electrons and holes are not easy to measure experimentally. However, Monte Carlo calculations based on quantum mechanical scattering amplitudes gives a good idea of these parameters. The saturation velocities obtained from Monte Carlo simulations usually give good agreement with many device characteristics, for example, the high electron mobility transistor (HEMT). Typical saturation velocities for the group III nitrides can range from 2×10^7 cm sec^{-1} to 5×10^7 cm sec^{-1}. It is possible to obtain a near-terahertz or terahertz response using semiconductor layers thinner than 100 nm.

The steady-state current can be affected by carrier multiplication through impact ionization. However, as the semiconductor layer is made thinner to obtain a faster response, the amount of impact ionization goes down drastically. Also, for ultra-thin MSM detectors, the electrons and holes can transit the semiconductor without fully reaching bulk-like steady states. This is the regime of transient velocity overshoot, and it can substantially increase the response speeds of these devices.

The MSM structure is typically obtained using a lateral MSM spacing. This means that the type of lithography used to define the MSM structure is relevant to the kind of speeds that can be obtained. Plasmonic enhancement through metallic nanoparticles on nitride semiconductors has been used to obtain an enhanced photoresponse in GaN and related MSM photodetectors, as shown in Figures 7.5 and 7.6. Plasmons in nanoparticles modify the electromagnetic field profile as well as the band profile in the semiconductor, which leads to better absorption and photoresponse.

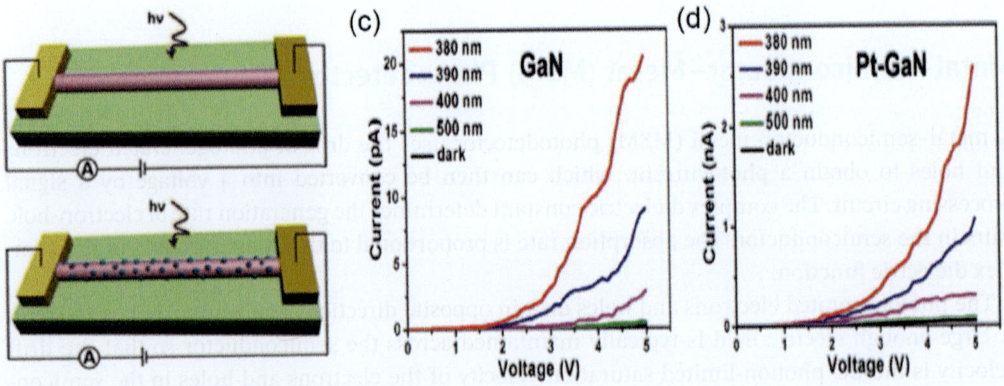

Figure 7.5 GaN photodetector and platinum (Pt)-GaN plasmonic enhancement photodetector characteristics. From Reference [5] – used with permission.

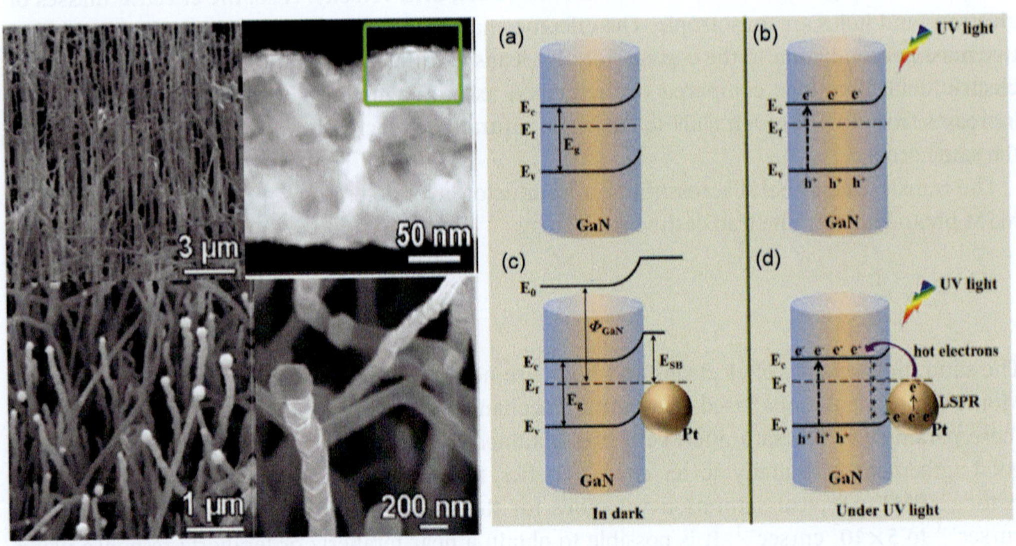

Figure 7.6 Pt-GaN plasmonic enhancement photodetector cross-sections and band-diagrams. From Reference [5] – used with permission.

Asymmetric MSM-type photodetectors have also been demonstrated. For example, graphene has been used as one electrode with metal layers as another electrode to obtain a good photoresponse. Graphene/nanoporous GaN photodetectors with improved photoresponses have been demonstrated. The nanoporous GaN leads to better light confinement and absorption properties. GaN/molybdenum diselenide ($MoSe_2$) heterostructure photodetectors have been used to obtain a broad spectrum photoresponse. This heterojunction uses the type-II heterojunction between GaN and $MoSe_2$ to obtain a good transport of photogenerated carriers. The choice of metals and the Schottky barrier heights to the semiconductor layers in these structures crucially affects device characteristics.

Since the MSM photodetector uses a voltage bias to sweep away photogenerated carriers, there is a dark current when the signal is absent. The total current can be written as the sum

$$I_{\text{total}} = I_{\text{photo}} + I_{\text{dark}} \tag{7.11}$$

The shot noise power spectral density in steady state is given by

$$I_{shot}(j\omega) = 2qI_{total} \tag{7.12}$$

The photogenerated current typically passes through a resistor to convert the current to a voltage signal. The resistor has Nyquist noise associated with it and has a power spectral density given by

$$V_R(\mathrm{j}\omega) = \frac{4kT}{R} \tag{7.13}$$

where k is Boltzmann's constant, T is the temperature, and R is the resistance. The signal-to-noise ratio (SNR) is given by

$$SNR = \frac{I_{photo}R}{\int_0^{\infty}\left(I_{shot}\left(j\omega\right)R + \frac{4kT}{R}\right)H(j\omega)d\omega} \tag{7.14}$$

where $H(\mathrm{j}\omega)$ is an effective transfer function that can be calculated from the small-signal circuit of the photodetector and load. SNR maximization requires that the dark current should be small compared to the photogenerated current. The dark current in the MSM detector is the tunneling-assisted thermionic emission current or the thermionic-field current. This requires choosing the metal Schottky barrier height carefully to obtain lower tunneling and lower thermionic emission. The dark current can be written as

$$\beta(E) = \beta_0 \exp\left(-\frac{E_{\beta 0}}{E}\right) \tag{7.15}$$

where $f(k_x,k_y,k_z)$ is the Fermi–Dirac distribution function of electrons, T is the quantum mechanical transmission coefficient, and $k_x k_y$, and k_z are the crystal momentum components along the x-, y-, and z-axes, respectively. The transmission coefficient can be calculated from a numerical solution to the Schrodinger equation. In many cases, the Wentzel–Kramers–Brillouin (WKB) approximation or transfer-matrix methods can be used to obtain the transmission coefficient. It decreases steeply as the barrier height increases or as the barrier width increases. It also decreases with the effective mass of the tunneling electrons. InN has an effective mass of 0.08, GaN an effective mass of about 0.25, and AlN has an effective mass of about 0.4 times the free electron mass. This means that for the same width and barrier height, InGaN tends to have a higher transmission coefficient compared to AlGaN. Transmission coefficients and dark currents should be calculated for each specific design so that SNR optimization can be done.

High-temperature MSM photodetection can be achieved by device optimization. The Schottky barrier height at the metal–semiconductor junction is crucial to achieving low dark currents and good responsivity.

Solar-blind Group III-Nitride UV Photodetectors

The bandgap of GaN is 3.43 eV, which means that it absorbs photons only in the UV range or higher energies. This makes it possible to obtain UV photodetectors that are solar blind. AlGaN can take the bandgap even higher and solar-blind detectors become easier to obtain.

The device dimensions required to obtain good solar-blind detection depend on the UV cut-off that the device tries to attain. As the cut-off value of energy (or frequency) goes up, the absorption coefficient goes down. This means that thicker device layers are required to obtain the desired photoresponse and required SNRs.

The effective mass of electrons and holes increases as the Al content increases in AlGaN. The saturation velocity typically decreases with effective mass of carriers with a $tan(h[\hbar\omega/m^*])$ dependence, approximately. This means that as one designs solar-blind UV detectors using the MSM structure, for example, the speed decreases as one moves deeper into the UV regime.

One major microscopic process that occurs when photodetectors are used to detect above bandgap radiation is that the photogenerated carriers relax to near the valence-band maximum or conduction-band minimum by emitting phonons (losing energy to phonons). This decides if the relevant transport processes like drift are those applicable to near-band maxima models or to some more complicated full-band model. In typical photodetector designs, the former is the case so that complicated full-band phenomena do not play too important a role. The reader can find more details about group III-Nitride UV photodetectors in References [7–37].

p-i-n Photodiodes

Adding an intrinsic layer between the p and the n layers provides a larger photon absorbing layer for the same maximum electric field, as shown in Figure 7.7. This *p-i-n* design can improve the efficiency over a *p-n* diode.

The generation rate of electrons and holes is given by

$$G(x) = (1 - R)\frac{P_{opt}}{h\nu}\alpha e^{-\alpha x} \tag{7.16}$$

where distance x is measured from the side from which the optical radiation is incident. The total current at any location is the sum of diffusion and drift currents.

$$J_{total} = J_{diffusion} + J_{drift} \tag{7.17}$$

The continuity equations for electrons and holes can be written similarly to Equation 7.1 and the net photocurrent can be obtained. The electron and hole generation in the intrinsic drift layer controls the overall photoresponse in good designs.

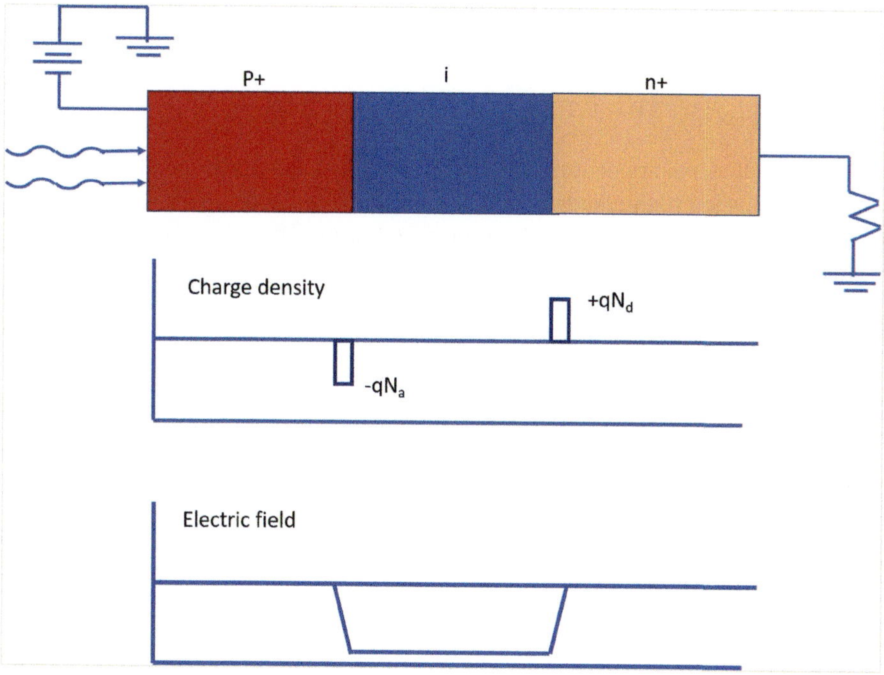

Figure 7.7 Schematic, charge density profile, and electric field profile of a *p-i-n* photodiode.

Avalanche Photodiodes

There is a lot of interest in obtaining photodetection of low-intensity radiation with large SNRs. For example, single-photon detection has become a topic with vast interest because it opens the way to sophisticated computation using single photons in various superposition states. The wide bandgap of the nitrides means that this kind of detection for high-energy photons (especially in the UV range) can be achievable using avalanching effects in avalanche photodiodes (APDs).

$$\frac{d}{dx}J_n(x)= \alpha_n(x)J_n(x)+\beta_p(x)J_p(x)+\mathrm{qG}(x) \tag{7.18}$$

$$-\frac{d}{dx}J_p(x)= \alpha_n(x)J_n(x)+\beta_p(x)J_p(x)+\mathrm{qG}(x) \tag{7.19}$$

$$J=\frac{qG_0e^{\int_x^W[\alpha_n(x')-\beta_p(x')]dx'}}{1-\int_0^W dx'^{\beta_p}(x')e^{\int_{x'}^W[\alpha_n(x'')-\beta_p(x'')]dx''}} \tag{7.20}$$

$$\langle i_S^2\rangle=2q(I_P+I_B+I_D)\langle M^2\rangle \tag{7.21}$$

The SNR of the APD receiver circuit is given by

$$SNR = \frac{\frac{1}{2}\left(\frac{q\eta P_{opt}}{h\nu}\right)^2 \langle M\rangle^2}{2q(I_P + I_B + I_D)\langle M\rangle^2 \mathrm{FB} + 4kT / R_{eq}} \tag{7.22}$$

The impact ionization coefficients are determined by the rates of generation of conduction-band holes due to collisions between high-energy conduction-band electrons and valence-band electrons. The electron in the conduction band must have energy (measured from bottom of conduction band) higher than the bandgap energy to initiate impact ionization. The actual threshold for impact ionization can be higher depending on the requirements of energy conservation and crystal momentum conservation. The same holds for an impact ionization event initiated by holes. Figure 7.8 shows a schematic diagram for the microscopic impact ionization process.

In steady states, the impact ionization coefficients depend on electric fields through exponential functions.

$$\alpha(E) = \alpha_0 \exp\left(-\frac{E_{\alpha 0}}{E}\right) \tag{7.23}$$

$$\beta(E) = \beta_0 \exp\left(-\frac{E_{\beta 0}}{E}\right) \tag{7.24}$$

Also, the three kinds of holes (heavy, light, and spin-orbit split-off holes) have different impact ionization coefficients due their different effective masses.

The response speed of the *p-i-n* avalanche photodiode depends crucially on transit time across the intrinsic layer. For devices optimized for speed, the transport of electrons and holes can be in the non-local or transient velocity overshoot regime. In that case, the impact ionization coefficient can be different from the simple forms above and Monte Carlo calculations can become necessary to obtain the impact ionization characteristics. Ref [6] provides experimental values of impact ionization coefficient in GaN.

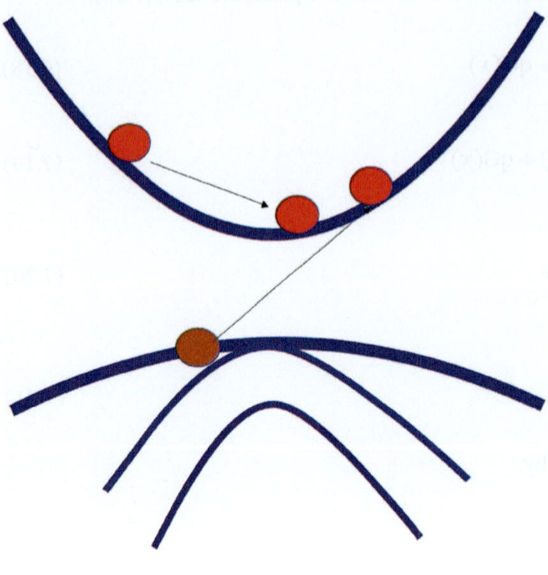

Figure 7.8 Impact ionization processes and carrier multiplication showing the conduction band and three valence bands (heavy-hole band, light-hole band, and spin-orbit split-off band).

At sufficiently high electric fields, band-to-band tunneling can also contribute to ionization processes. This kind of tunneling transfers a valence-band electron to the conduction band through quantum mechanical tunneling. The tunneling amplitudes are sensitive to the barrier height, which increases as the bandgap increases. This means that these processes are more likely to affect device characteristics in InGaN compared to AlGaN. Furthermore, the heavy-hole band, the light-hole band, and the spin-orbit split-off band contribute differently to band-to-band tunneling.

The impact ionization coefficients tend to be low in the group III-Nitride material system for the same electric field compared to the arsenide or the phosphide material systems. Therefore, *p-i-n* avalanche photodiodes need to be biased with

sufficient reverse bias to get the required level of carrier multiplication through impact ionization. Figure 7.9 shows the experimentally determined impact ionization coefficients for electrons and holes in GaN, according to one study. It must be kept in mind that the bandgap of AlGaN alloy is larger than that of GaN, and therefore, the impact ionization coefficient is also lower compared to GaN. Similarly, in the InGaN system, the impact ionization coefficient decreases as one increases the gallium concentration. Heterojunction photodiode is another class of devices of interest in group III nitrides. They involve real-space transfer of electrons and/or holes, as shown in Figure 7.10.

There is a lot of interest in achieving single-photon detection with high SNRs using avalanche photodiodes. These photodetectors are called single-photon avalanche diodes (SPADs). In the blue and ultraviolet wavelengths, there is interest in using nitride semiconductors to achieve single-photon detection. It must be kept in mind that achieving carrier multiplication for single-photon detection requires large electric fields in the nitride semiconductor material system. This is because impact ionization coefficients as a function of the electric field tend to be low in these wide band-gap materials. The voltage required to achieve these high fields must be carefully calibrated for any specific detector design because the breakdown voltage is a function of the entire electric field profile inside the device.

The carrier multiplication or avalanching due to impact ionization can potentially be offset by recombination at non-radiative recombination sites like dislocations. This means that achieving high-quality and high-performance SPADs requires minimizing the dislocation density in the detector layers. Since the dislocation density is a function of substrate and growth conditions, the

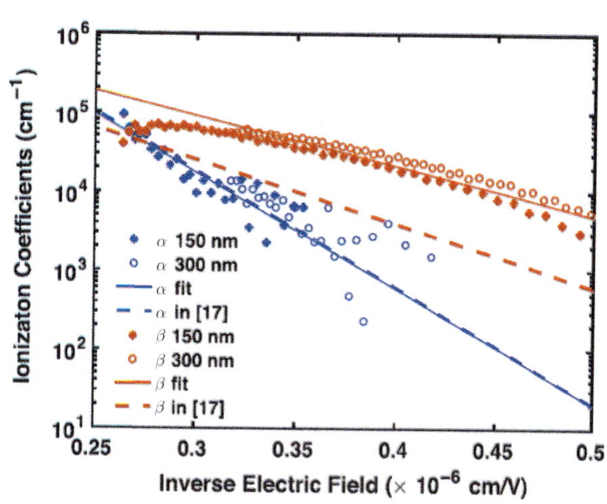

Figure 7.9 Experimental impact ionization coefficients in GaN as a function of the inverse electric field. From Reference [38], used with permission.

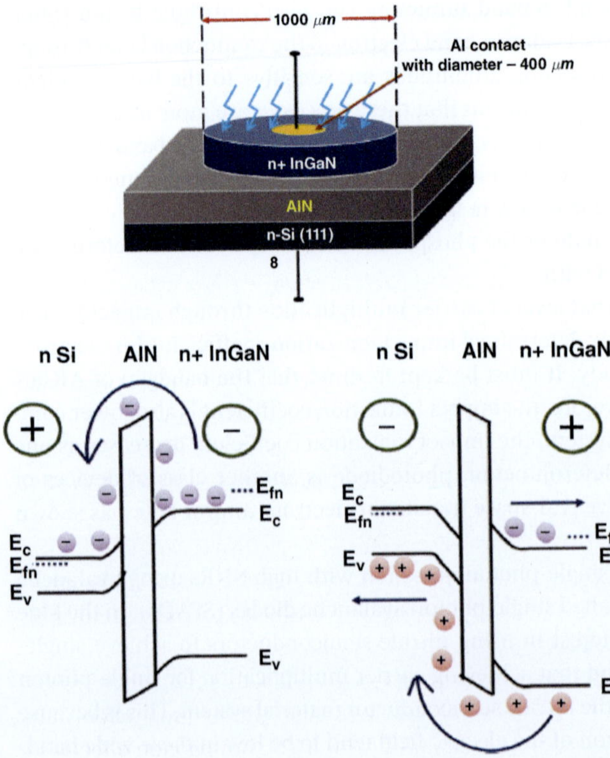

Figure 7.10 Heterojunction InGaN/AlN/Si photodetector. From Reference [39] – used with permission.

optimization of substrate-device interactions and of growth conditions becomes crucial to achieving detector device layers with the right characteristics.

As the use of SPADs increases, there is interest in achieving high-speed SPADs that can detect many single photons separated by small time intervals. This requires optimizing the carrier multiplication rates and the drift speed in the device layers as well as the microscopic speed of the avalanche process itself. While there have been several theoretical and experimental studies of the steady-state impact ionization coefficients, far less attention has been given to the intrinsic speeds of these processes. In impact ionization, a high-energy electron loses some energy, and this energy is used to excite an electron from the valence band to the conduction band. This process has an intrinsic speed, and this speed varies from one material system to another. For example, this speed, typically given by some kind of quantum mechanical Fermi's golden rule kind of calculation, is different from gallium arsenide (GaAs) to GaN. And it is also different from GaN to AlGaN. A better understanding of the time scales involved in these processes can help design future high-speed SPADs.

The impact ionization process involves energetic electrons, which also lose a lot of energy to the lattice and heat up the device layers. This can lead to hot phonon effects, which refer to situations where the occupation number of phonons become a non-equilibrium one so that the impact ionization process depends on this hot phonon distribution. The impact ionization coefficients calculated or measured for bulk layers may become less relevant in the photodetector device because the phonons lose their energy to the substrate and the surroundings differently in bulk materials and in devices. The actual impact ionization coefficients can, therefore, vary quite a bit from bulk materials to devices. Careful quantification becomes of the essence if high-performance SPADs or similar single-photon detectors are to be achieved. Advanced simulation methods, like Monte Carlo simulations that consider non-equilibrium phonon distributions, may become necessary to understand these effects in advanced device structures. The requirement for optoelectronic integration and advanced scaling can make this hot phonon effect a major effect to be understood and quantified to achieve desired device goals.

Schottky Barrier Photodiodes

Metal–semiconductor Schottky barriers under reverse bias can be used for photodetection with proper design based on parameters like Schottky barrier height, doping density in the semiconductor, and carrier recombination lifetimes. Schottky barrier diodes can also be biased in the carrier multiplication region to get the avalanche effect, with SNR optimization similar to that explained for *p-i-n* avalanche photodiodes.

Schottky barrier photodiodes can potentially lead to better immunity against dislocation effects because they offer different requirements on device layer widths. However, the exact optimization between dislocation densities and absorption coefficients in the various device layers needs careful calculation.

Schottky barrier photodiodes can also potentially be designed to achieve single-photon detection. Once again, understanding the trade-offs between the impact ionization coefficient and device layer widths and depletion widths needs careful quantification. The hot phonon effect also needs to be quantified carefully, keeping the actual device layer design in mind, so that actual impact ionization coefficients can be derived.

The Schottky barrier height is another device parameter that must be chosen carefully because the dark current (current flowing in the absence of incident radiation) depends sensitively on the Schottky barrier height. This is because when the Schottky barrier photodiode is reverse biased, the dark current flows due to quantum mechanical tunneling between the semiconductor *n*-layer and the Schottky metal electrode. The quantum mechanical tunneling probabilities are strongly influenced by barrier heights and by electron effective mass. The tunneling barrier height depends on the Schottky barrier height, and therefore, the latter becomes an important parameter that must be considered in any specific design.

Heterogenous Photodiodes with Group III Nitrides and Transition Metal Dichalcogenides

Transition metal dichalcogenides (TMD)-group III nitride heterojunction photodiodes use strain effects to optimize performance metrics like speed and signal-to-noise ratios. Strain affects the band structure in the TMD and to some extent in the group III nitride. This has impact on mobility and saturation drift velocity in these materials as well as on photodetector performance. Conduction-band offsets are crucial parameters for heterogenous photodetectors.

Heterojunction photodiodes using nitrides and TMDs do involve more complex growth processes because a wide variety of materials and precursor gases need to be used to grow these heterostructures. Therefore, beyond research reports, the design of these heterogenous photodiodes for commercialization or specialized uses involves a lot of cost optimization. However, epitaxial growth of semiconductors is also rapidly evolving, and speed and costs tend to come down with time. Therefore, these heterogenous photodiodes may become more accessible with time.

Achieving single-photon detectors using heterogenous photodiodes will obviously involve using large electric fields and impact ionization. It must be kept in mind that impact ionization near heterojunctions can be very different from those in bulk materials. So, designing single-photon detectors using these devices requires an understanding of impact ionization processes in heterojunction settings.

Another physical effect that needs careful quantification for these photodiodes is the interface charge at the dichalcogenide-nitride heterojunction due to polarization discontinuities. This is because the group II nitrides have spontaneous and piezoelectric polarization especially for *c*-axis growth. This can lead to band profile changes and overall changes in the device performance like SNR and speed. The reader can find details about heterojunction and heterogenous group III-Nitride photodetectors in References [40–50].

Alloy Nitrides and Spectral Response

The spectral range of infrared through visible to the UV can be spanned by InGaN. The range for UV to deep UV can be spanned by AlGaN. Alloy scattering plays a major role in low field mobilities in these materials. However, the saturation velocity is not affected a lot by alloy disorder because alloy scattering is typically low at larger energies.

For those photodetectors that rely on low field mobilities, large levels of alloying can affect detector efficiency because there can be significant carrier localization with corresponding lower mobilities, and carriers can recombine more before being swept to the contacts. Growth conditions can be critical to deciding the amount of clustering in to allow and impact on mobilities and recombination times. There can be subtle effects on device efficiency in MSM detectors as well when the alloying level is high because hopping conduction can play a role when carrier localization is significant.

Furthermore, the fundamental question of how dislocation densities in alloys with large alloying levels depends on substrate and growth conditions needs careful quantification. So far, the reports in the literature on this subject are scanty and so there is a lot of scope for improving the understanding of dislocations and their effect on detector performance when alloys with high alloy disorder are used as device layers.

Polarization and Photodetector Performance

Since group III-Nitride semiconductors have large spontaneous and piezoelectric polarization, the question arises as to how much these polarizations affect photodetector properties. For MSM photodetectors grown along the *c*-axis and using drift transport perpendicular to the *c*-axis, the built-in electric fields along the *c*-axis can cause some drift perpendicular to the main direction of transport. For typical MSM photodetector designs, this kind of effect causes only a small change in overall device characteristics. However, as more scaled photodetectors are designed and produced, the spontaneous and piezoelectric polarization effects can become non-negligible in advanced designs.

For *p-i-n* photodetectors and *p-i-n* avalanche photodiodes, polarization grading offers an added design dimension to tailor band profiles and electric fields across the device. Polarization grading can be used to obtain bound charge densities of specific sign, which in turn can modify the band profile. Therefore, the impact of ionization coefficients as a function of distance can be modified. It must be kept in mind, however, that polarization grading using composition grading makes the growth process more complicated, whether it is molecular beam epitaxy (MBE) or metal-organic chemical vapor deposition (MOCVD). Therefore, polarization grading tricks can only be used when high performance requires it and when process cost and process time are not the major considerations.

For Schottky barrier photodetectors again, polarization grading can potentially lead to another device design dimension. However, these photodetectors are typically used for decreased complexity of growth and fabrication. So, polarization tricks are less amenable to these photodetectors.

Future photodetector designs that involve aggressive scaling for specific applications may need to contend with polarization, not just as an added design dimension but also as a potential problem for detector performance. *m*-Plane and *a*-plane growth may become necessary for highly specialized designs to eliminate unnecessary polarization effects. It must be kept in mind that these *m*-plane and *a*-plane designs can come with their own problems like high dislocation densities and high stacking fault densities. So, careful device and process optimization become necessary.

Advanced detectors using second photon absorption with quantum dots are potentially manufacturable and can have strong dependence on polarization effects. However, at this stage, it is not clear what kind of applications this can lead to better detectors for. Therefore, this can only be a subject of research explorations and quantification for now.

Photodetectors and Substrate Engineering

The photogenerated carriers need to be collected swiftly and efficiently at the electrodes for optimum photodetector operation. It must be kept in mind that photogenerated carriers can undergo various types of recombination like band-to-band recombination and Shockley–Read–Hall (SRH) recombination. The latter type of recombination is mediated by mid-gap traps and its rate increases as the density of traps increases. For nitride semiconductors, these traps can be due to dislocations and vacancies. The density of dislocations is sensitive to the growth conditions and to the relative lattice mismatch between the epilayers and the substrate as well as the relative lattice mismatch between different epilayers if the design consists of more than one epilayer.

Since cost optimization is a vital part of photodetector design, as in any engineering design, the use of cheap substrates like silicon and sapphire becomes relevant in many cases. The relatively large lattice mismatch between GaN and these substrates means that the dislocation density becomes a parameter of concern. The use of specialized templates with dislocation-absorbing layers is of interest in many photodetector applications. The drift time of photogenerated carriers in *p-i-n* or MSM photodetectors must be much smaller than the overall recombination time, which is determined by the sum of the band-to-band and SRH recombination rates. Also, too large a dislocation density can lower the saturation velocity of electrons and holes due to extra scattering at these dislocations. This can lead to higher drift times for photodetectors. This not only reduces the photodetector speed, but also decreases its quantum efficiency.

The quantum efficiency of the photodetector can be written as the ratio of the number of charge carriers collected per second to the number of photons absorbed per second. Clearly, higher dislocation density leads to lower quantum efficiency because many of the carriers will recombine non-radiatively at these dislocations, which act as non-radiative recombination centers.

For high lattice-mismatch substrates like sapphire and silicon, one way to reduce dislocations in the device layer is to grade the composition so that lower lattice-mismatch layers are closer to the substrate to absorb dislocations so that the device layers are relatively dislocation-free. This obviously increases process complexity because a large number of layers must be grown before reaching the device layers, each with its own optimum growth conditions like temperature and flow rate. Highly specialized photodetectors with high quantum efficiency requirements can potentially be built on more expensive substrates like GaN and silicon carbide (SiC), which offer much smaller lattice mismatches and much lower dislocation densities. The cost of these kinds of substrates is high, leading to the obvious performance-cost trade-offs.

Using multi-layer templates with a superlattice-type structure can be a useful way of absorbing dislocations in template layers so that the device layers are free of dislocations. These kinds of template growth methods become relevant when high performance is needed and when process cost and time are less constraining. This is a subject of ongoing research and can lead to device improvements with time.

References

1 Feneberg, M. et al. (2012). Optical properties of cubic GaN from 1 to 20eV. *Phys. Rev. B* 85 (155207).
2 Djurisic, A. et al. (1999). Modeling the optical constants of hexagonal GaN, InN and AlN. *Appl. Phys. Lett.* 85 (5).
3 Djurisic, A. et al. (2001). Dielectric function models for describing the optical properties of hexagonal GaN. *J. Appl. Phys.* 89 (1).
4 Guo, D. et al. (2018). Self-Powered Ultraviolet Photodetector withSuperhigh Photoresponsivity (3.05 A/W) Based on the GaN/Sn: ga2O3pn Junction. *ACS Nano* 12: 12827–12835.
5 Zhang, X. et al. (2017). Giant UV photoresponse of a GaN nanowire photodetector through effective Pt nanoparticle coupling. *J. Mater. Chem. C* 5: 4319.
6 Cao, L. et al. (2018). Experimental characterization of impact ionization coefficients for electrons and holes in GaN grown on bulk GaN substrates. *Appl. Phys. Letts* 112.
7 Kang, C. et al. (2020). Semipolar (20$\bar{2}\bar{1}$) InGaN/GaN micro-photodetector for gigabit-per-second visible light communication. *Appl. Phys. Exp.* 13: 014001.
8 Chen, Y. et al. (2021). A high-response ultraviolet photodetector by integrating GaN nanoparticles with graphene. *J. Alloys Compounds* 868.
9 Li, Z. et al. (2020 June). Strain enhancement for a MoS2-on-GaN photodetector with an Al_2O_3 stress liner grown by atomic layer deposition. *Photonics Res.* 8 (6).
10 Sanjay, S. et al. (2020). Epitaxy of gallium nitride pyramids on few layer graphene for metal-semiconductor-metal based photodetectors. *Mat. Chem. Phys.* 240.
11 Zhuo, R., Wang, Y., Wu, D. et al. (2018). High-performance self-powered deep ultraviolet photodetector based on MoS2/GaN p–n heterojunction. *J. Mater. Chem. C* 2.
12 Yu, R., Wang, G., Shao, Y. et al. (2019). From bulk to porous GaN crystal: precise structural control and its application in ultraviolet photodetectors. *J. Mater. Chem. C* 45.
13 Müller, A., Konstantinidis, G., Androulidaki, M. et al. (2012). Front and backside-illuminated GaN/Si based metal– semiconductor–metal ultraviolet photodetectors manufactured using micromachining and nano-lithographic technologies. *Thin Solid Films* 520 (6).
14 Wang, F. (2017). Theoretical study of back-illuminated separated absorption and multiplication AlGaN APDs with different structural parameters. *Opt. Quant. Elect.* 49.
15 Li, K.H., Fu, W.Y., Cheung, Y.F. et al. (2018). Monolithically integrated InGaN/GaN light-emitting diodes, photodetectors, and waveguides on Si Substrate. *Optica* 5 (5).

16 Jeong, H., Bang, S., Oh, H.M. et al. (2015). Semiconductor-insulator-semiconductor diode consisting of monolayer MoS2, h-BN, and GaN heterostructure. *ACS Nano* 9 (10).

17 Goswami, L., Pandey, R., and Gupta, G. (2020). Ultra-thin GaN nanostructures based self-powered ultraviolet photodetector via nonhomogeneous Au-GaN interfaces. *Opt. Mater.* 102.

18 Tchernycheva, M., Messanvi, A., de Luna Bugallo, A. et al. (2014). Integrated photonic platform based on InGaN/GaN nanowire emitters and detectors. *Nano Lett.* 14 (6).

19 Dubey, A. et al. (2020). Aluminum plasmonics enriched ultraviolet GaN photodetector with ultrahigh responsivity, detectivity, and broad bandwidth. *Adv. Sci.* 7 (24).

20 Mou, W., Zhao, L., Chen, L. et al. (2017). GaNbased Schottky barrier ultraviolet photodetectors with graded doping on patterned sapphire substrates. *Solid-State Electron* 133.

21 Pandit, B. et al. (2020). Dual-functional ultraviolet photodetector with graphene electrodes onAlGaN/GaN heterostructure. *Sci Rep.* 10 ScS.

22 Jain, S.K. et al. (2021). 2D/3D hybrid of MoS_2/GaN for a High-performance broadband photodetector. *ACS Appl. Electron. Mater.* 3 (5).

23 Liu, C. et al. (2021). A nanopillar-modified high-sensitivity asymmetric graphene–GaN photodetector. *Nanoscale* 41.

24 Zhang, M., Wang, Y., Teng, F. et al. (2016). A photoelectrochemical type self-powered ultraviolet photodetector based on GaN porous films. *Mat. Lett.* 162.

25 Hu, A. et al. (2019). Graphene on Self-assembled InGaN quantum dots enabling ultrahighly sensitive photodetectors. *Adv. Opt. Mat.* 7 (8): 1801792.

26 Kasic, A. et al. (2004). InN dielectric function from the midinfrared to the ultraviolet range. *Phys. Rev. B* 70: 115217.

27 Wang, G., Lu, H., Chen, D. et al. (2013). High quantum efficiency GaN-based p-i-n ultraviolet photodetectors prepared on patterned sapphire substrates. *IEEE Photon. Technol. Lett.* 25 (7).

28 Liu, L., Yang, C., Patanè, A. et al. (2017). High-detectivity ultraviolet photodetectors based on laterally mesoporous GaN. *Nanoscale* 24.

29 Ho, K.-T., Chen, R., Liu, G. et al. (2018). 32 Gigabitper- second visible light communication link with InGaN/GaN MQW Micro-photodetector. *Opt. Exp.* 26 (3).

30 Gundimeda, A., Krishna, S., Aggarwal, N. et al. (2017). Fabrication of non-polar GaN based highly responsive and fast UV photodetector. *Appl. Phys. Lett.* 110 (10).

31 Yoshikawa, A., Ushida, S., Nagase, K. et al. (2017). High-performance solar-blind Al0.6Ga0.4N/ Al0.5Ga0.5N MSM type photodetector. *Appl. Phys. Lett.* 111 (19).

32 Sakalauskas, E. et al. (2011). Dielectric function and optical properties of quaternary AlInGaN alloys. *J. Appl. Phys.* 110: 013102.

33 Li, D., Jiang, K., Sun, X., and Guo, C. (2018). AlGaN photonics: recent advances in materials and ultraviolet devices. *Adv. Opt. Photonics* 10 (1).

34 Peng, M., Liu, Y., Yu, A. et al. (2016). Flexible Self-powered gan ultraviolet photoswitch with Piezo-phototronic effect enhanced On/Off ratio. *ACS Nano* 10 (1).

35 Lee, J.-H., Lee, W.W., Yang, D.W. et al. (2018). Anomalous photovoltaic response of graphene-on-GaN Schottky photodiodes. *ACS Appl. Mater. Interfaces* 10 (16).

36 Sun, X., Li, D., Jiang, H. et al. (2011). Improved performance of GaN metal-semiconductor-metal ultraviolet detectors by depositing SiO2 nanoparticles on a GaN surface. *Appl. Phys. Lett.* 98 (12).

37 Chang, Y.H., Hsu, T.C. et al. (2021). High-bandwidth InGaN/GaN semipolar micro-LED acting as a fast photodetector for visible light communications. *Opt. Exp.* 29 (23).

38 Chowdhury, A. et al. (2019). Self-powered, broad band, and ultrafast InGaN-based photodetector. *ACS Appl. Mater. Interfaces* 11: 10418–10425.

39 Li, J. et al. (2020). Ultrahigh sensitivity graphene/Nanoporous GaN ultraviolet photodetectors. *ACS Appl. Mater. Interfaces* 12: 11965–11971.

40 Ho, C.H., Lin, G.J., Fu, P.H. et al. (2012). An Efficient Light-Harvesting Scheme Using SiO2 Nanorods for InGaN Multiple Quantum Well Solar Cells. *Sol. Energy Mater. Sol. Cells* 103.

41 Krishna, S., Sharma, A., Aggarwal, N. et al. (2017). Ultrafast photoresponse and enhanced photoresponsivity of Indium Nitride based broad band photodetector. *Sol. Energy Mater. Sol. Cells* 172.

42 Sadi, T. et al. (2014). On the correlation of the Auger generated hot electron emission and efficiency droop in III-N light-emitting diodes. *Appl. Phys. Lett.* 105.

43 Chang, S., Chang, M., and Yang, Y. (2017). Enhanced responsivity of GaN metal–semiconductor–metal (MSM) photodetectors on GaN substrate. *IEEE Photon. J* 9 (2).

44 Wang, J. et al. (2021 May). Polarization assisted self-powered GaN-based UV photodetector with high responsivity. *Photonics Res.* 9 (5).

45 He, T., Zhang, X., Ding, X. et al. (2019). Broadband ultraviolet photodetector based on vertical Ga2O3/GaN nanowire array with high responsivity. *Adv. Opt. Mater.* 7 (7).

46 Tang, X. et al. (2021). Temperature enhanced responsivity and speed in an AlGaN/GaN metal-heterostructure-metal photodetector. *Appl. Phys. Lett.* 119 (1).

47 Elbar, M., Alshehri, B., Tobbeche, S., and Dogheche, E. (2018). Design and simulation of InGaN/GaN p-i-n photodiodes. *Phys. Status Solidi A* 215 (9).

48 Goswami, L. et al. (2020). Ultra-thin GaN nanostructures based self-powered ultraviolet photodetector via non-homogeneous Au-GaN interfaces. *Opt. Mat.* 102.

49 Yi, C. et al. (2021). MXene-GaN van der Waals heterostructures for High-speed self-driven photodetectors and light-emitting diodes. *J. Electron. Mater.* 7: 5.

50 Pant, R., Shetty, A., Chandan, G. et al. (2018). In-Plane anisotropic photoconduction in nonpolar epitaxial a-Plane GaN. *ACS Appl. Mater. Interfaces* 10 (19).

8

Photovoltaics and Energy Conversion Devices

Indium Gallium Nitride Material System for Solar Cells

Since the indium gallium nitride (InGaN) system covers the wavelength range from infrared to ultraviolet (UV) through the visible, it is possible to design photovoltaic cells in this material system for good power-conversion efficiencies and quantum efficiencies. One other optimal feature of the InGaN alloy system is that these alloys are direct bandgap in nature, with the top of the valence bands occurring at the same position as the bottom of the conduction band in the Brillouin zone. This is in contrast to many other technologically common semiconductor systems like the arsenides or the phosphides where the UV and high energy visible parts of the spectrum are absorbed with low overall device efficiency because the absorption is well above bandgap and hot carriers lose energy to lattice vibrations, especially longitudinal optical (LO) phonons.

Basic Solar Cell Physics – *p-n* Junction Solar Cells

We begin with the physics of *p-n* junction photovoltaic cells. The ionization of donors and acceptors in the *p-n* junction device is given by

$$N_d^+ = \frac{N_d}{1 + g_D e^{\left(\frac{E_f - E_d}{kT}\right)}} \tag{8.1}$$

and

$$N_a^- = \frac{N_a}{1 + g_A e^{\left(\frac{E_a - E_f}{kT}\right)}} \tag{8.2}$$

For direct bandgap semiconductors, the light absorption is given by the absorption coefficient, which depends on the photon energy as

$$\alpha(h\nu) = A^*(h\nu - E_g)^{1/2} \tag{8.3}$$

for photon energies not too far above the bandgap energy. For energies well above the bandgap energy, the dependence of absorption coefficient on energy is more complex and reflects a more

Group III-Nitride Semiconductor Optoelectronics, First Edition. C. Jayant Praharaj.

complicated structure in conduction-band density of states (DOS) and valence-band DOS. The rate of creation of electron-hole pairs at a position x in the solar cell (where x is measured from the contact where the solar radiation is incident) is given by

$$G(x) = (1-s)\int_{\lambda=0}^{\lambda=\infty} (1-r(\lambda))f(\lambda)\alpha(\lambda)e^{-\alpha x}d\lambda \tag{8.4}$$

The equation above is valid when no composition grading is used. When composition grading is used, the absorption coefficient is itself a function of position and αx should be replaced by $\int \alpha(x)dx$ in the exponential. $f(\lambda)$ is the incident solar radiation spectrum. Figure 8.1 shows a typical solar radiation spectrum as a function of wavelength.

Recombination processes in the solar cell can be radiative recombination or non-radiative recombination. Radiative recombination processes are given by

$$R_{\text{radiative}} = B(pn - n_i^2) \tag{8.5}$$

Trap-mediated recombination of the Shockley–Read–Hall (SRH) type is given by

$$R_{SRH} = \frac{pn - n_i^2}{\tau_n\left(p + n_i e^{\frac{E_i - E_t}{kT}}\right) + \tau_p\left(n + n_i e^{\frac{E_t - E_i}{kT}}\right)} \tag{8.6}$$

The rate simplifies in p-type material to

$$R = \frac{n - n_0}{\tau_n} \tag{8.7}$$

and similarly for n-type material.

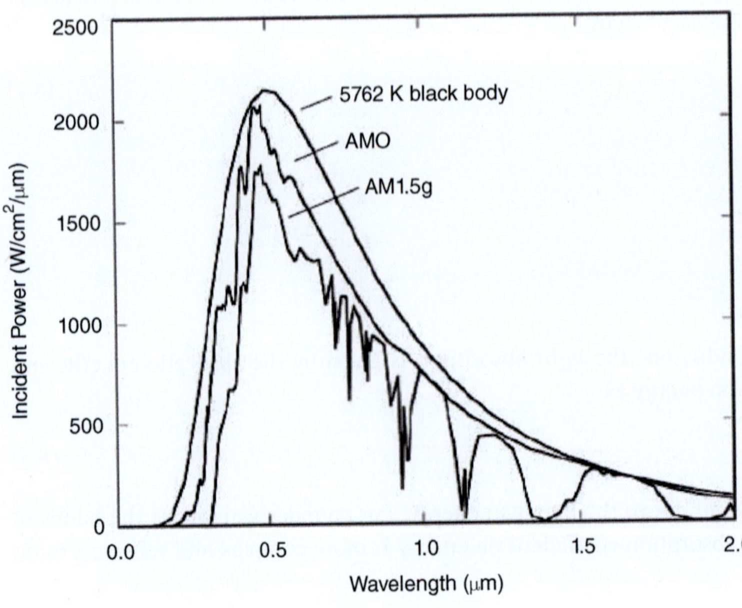

Figure 8.1 Nature of the solar radiation spectrum.

The drift and diffusion currents of electrons and holes are given by

$$\boldsymbol{J}_p^{drift} = q\mu_p p\boldsymbol{E} \tag{8.8}$$

where the bold-face $\boldsymbol{J}$ indicates that it is the current density vector.

$$\boldsymbol{J}_n^{drift} = q\mu_n n\boldsymbol{E} \tag{8.9}$$

$$\boldsymbol{J}_p^{drift} = -qD_p \nabla p \tag{8.10}$$

$$\boldsymbol{J}_n^{diff} = qD_n \nabla n \tag{8.11}$$

$$\boldsymbol{J}_n = J_n^{\text{drift}} + J_n^{\text{diff}} \tag{8.12}$$

$$\boldsymbol{J}_p = J_p^{\text{drift}} + J_p^{\text{diff}} \tag{8.13}$$

Gauss's equation determines the electric field profile from the charge density

$$\nabla.(\varepsilon \boldsymbol{E}) = q(p - n + N_d^+ - N_a^-) \tag{8.14}$$

The electron and hole continuity equations are given by

$$\nabla.\boldsymbol{J}_n = q\left(R_n - G + \frac{\partial n}{\partial t}\right) \tag{8.15}$$

$$\nabla.\boldsymbol{J}_p = q\left(G - R_p - \frac{\partial p}{\partial t}\right) \tag{8.16}$$

Inserting the carrier current equations into the continuity equations one gets, in the one-dimensional approximation and in steady state,

$$q\mu_p \frac{d}{dx} pE - qD_p \frac{d^2 p}{dx^2} = q(G - R) \tag{8.17}$$

and

$$q\mu_n \frac{d}{dx} nE + qD_n \frac{d^2 n}{dx^2} = q(R - G) \tag{8.18}$$

In quasi-neutral regions, the electric field can be neglected, and the equations above simplify to the minority carrier diffusion equations

$$D_p \frac{d^2 \Delta p_N}{dx^2} - \frac{\Delta p_N}{\tau_p} = -G(x) \tag{8.19}$$

and

$$D_n \frac{d^2 \Delta n_P}{dx^2} - \frac{\Delta n_P}{\tau_n} = -G(x) \tag{8.20}$$

for the minority carriers in the n-region and p-region, respectively. The depletion width is given by

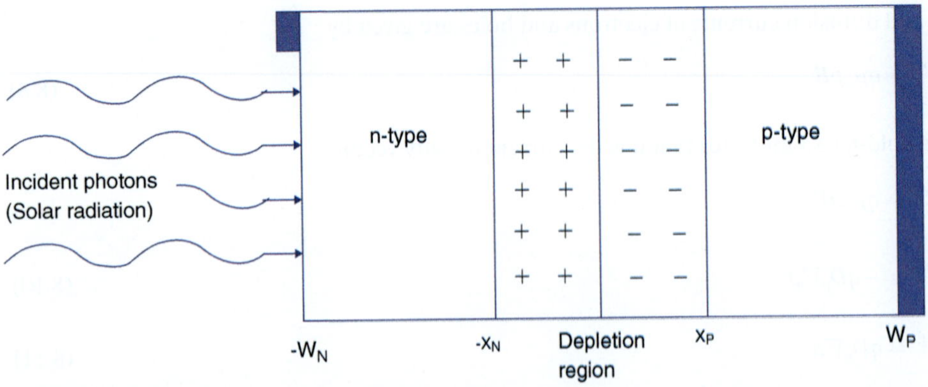

Figure 8.2 Schematic of *p-n* junction solar cell with depletion region.

$$W_D = x_N + x_P = \sqrt{\left(\frac{2\varepsilon}{q}\frac{N_a + N_d}{N_a N_d}\right) V_{bi}} \tag{8.21}$$

where V_{bi} is the built-in voltage of the *p-n* junction. Figure 8.2 shows a simple schematic of a *p-n* junction solar cell with depletion regions.

The boundary conditions at the front contact and back contact become crucial for determining the current-voltage characteristics of the photovoltaic cell.

$$D_p \frac{d\Delta p}{dx}\bigg|_{x=-W_N} = S_f \Delta p(-W_N) \tag{8.22}$$

where S_F is the surface recombination velocity at the front surface.

$$D_n \frac{d\Delta n}{dx}\bigg|_{x=W_P} = S_b \Delta n(W_P) \tag{8.23}$$

The assumption of flat quasi-Fermi levels inside the depletion region, which is usually good at low-level injection, leads to

$$p_N(-x_N) = \frac{n_i^2}{N_d} e^{\frac{qV}{kT}} \tag{8.24}$$

$$n_P(x_P) = \frac{n_i^2}{N_a} e^{\frac{qV}{kT}} \tag{8.25}$$

The minority carrier diffusion equations have solutions given by

$$\Delta p_N(x) = A\sinh\left(\frac{x + x_N}{L_P}\right) + B\cosh\left(\frac{x + x_N}{L_P}\right) + \Delta p_N'(x) \tag{8.26}$$

$$\Delta n_P(x) = C\sinh\left(\frac{x - x_P}{L_n}\right) + D\cosh\left(\frac{x - x_P}{L_n}\right) + \Delta n_P'(x) \tag{8.27}$$

where

$$V_{oc} = \frac{\eta kT}{q} \ln\left(\frac{J_{SC}}{J_D} + 1\right) \tag{8.28}$$

$$\Delta n_P'(x) = -(1-s)\int_\lambda \frac{\tau_N}{L_N^2\alpha^2 - 1}(1 - r(\lambda))f(\lambda)\alpha(\lambda)e^{-\alpha(x+W_N)}d\lambda \tag{8.29}$$

Integrating the continuity equation for electrons over the depletion region gives

$$\int_{-x_N}^{x_P} \frac{dJ_n}{dX} dX = J_n(x_p) - J_n(-x_n) = q\int_{-x_N}^{x_P} (R(x) - G(x))\mathrm{dx}$$

From this, we can derive [1],

$$I = A\left[J_p(-x_n) - J_n(x_p) + q\int_{-x_N}^{x_P} (G(x) - R(x))dx\right]$$

The overall *I–V* characteristics takes the form

$$I = I_{SC} - I_{01}\left(e^{\frac{qV}{kT}} - 1\right) - I_{02}\left(e^{\frac{qV}{2kT}} - 1\right)$$

The presence of the I_{SC} term is due to the generation term (absorption of solar radiation) and leads to negative current in positive bias (certain range of positive bias), which leads to power generation. Therefore, this photovoltaic cell can provide electrical power when illuminated by solar radiation. The optimization of the I_{SC}, I_{01}, and I_{02} terms leads to the optimization of power output from the photovoltaic cell. In this, material parameters like absorption coefficient, diffusion coefficient, and carrier lifetime play a major role. Device dimensions also play a major part in determining power output from the cell.

Quantum Efficiencies and Power-conversion Efficiencies

Out of all the electron-hole pairs generated every second by the absorption of photons from the incident solar radiation, only a certain fraction is collected as photocurrent in the solar cell. This is because recombination processes lead to a loss of conduction-band electrons and valence-band holes. The ratio of the number of electron-hole pairs actually collected as photocurrent or short-circuit current to the number of photons incident per second on the solar cell is called the quantum efficiency.

The power-conversion efficiency (PCE) is defined as

$$PCE = \frac{\textit{Maximum Electrical Power obtainable from Solar Cell}}{\textit{Incident Solar Optical Power}} \tag{8.30}$$

A typical form of the solar cell *I–V* characteristics is given approximately by

$$J = -J_{SC} + J_d\left(e^{\frac{qV}{\eta kT}} - 1\right) \tag{8.31}$$

where J_d is the dark current density of the solar cell, defined as the current that flows in the absence of solar radiation at high reverse bias. The open-circuit voltage of the solar cell under solar radiation can be found by equating the current density J to zero and is given by

$$V_{oc} = \frac{\eta kT}{q} \ln\left(\frac{J_{SC}}{J_D} + 1\right)$$

The maximum power density ($W\text{cm}^{-2}$) obtainable is derived by maximizing the absolute JV product with respect to V. The fill factor of the solar cell is defined as the ratio of the maximum power density to the product of J_{sc} and V_{oc}. Increasing the PCE requires optimizing both J_{sc} and V_{oc}. A large J_{sc} and small J_d are ideal for obtaining large PCEs.

p-i-n-Type solar cells can offer some advantages over *p-n* devices with proper design depending on material parameters. This is because *p-i-n* devices have a larger depletion layer and generation dominates over recombination in this region. With proper design, this can lead to a larger short-circuit current and higher PCE.

Effect of Composition Grading

Compositionally graded solar cells can help achieve energy loss to the lattice from photogenerated hot carriers by optimizing the absorption characteristics such that the high energy part of the solar spectrum gets absorbed by wider bandgap InGaN and the low energy part of the spectrum gets absorbed by lower bandgap InGaN. The composition grading can be of a discrete type, for example, a *p*-GaN/*n*-InGaN heterojunction, or it can be a continuously graded type. The discrete grading is easier to achieve technologically and is cost effective. The continuous grading is more difficult technologically but offers more scope for design optimization to improve power-conversion efficiencies. Reference [2] studies power-conversion efficiencies in *n*-InGaN/*p*-GaN single heterojunction solar cells and finds efficiency numbers ranging up to about 15%.

Discrete composition grading can take the form of quantum wells, multiple quantum wells (MQWs), and superlattices. The PCE in quantum well-based solar cells depends sensitively on the composition of the well layer, the composition of the barrier layer, width of the well layer, and overall device dimension parameters, like what portion of the device is covered by quantum wells and what portion is the barrier layer. Inserting a superlattice layer inside a barrier layer alters both the effective bandgap and the transport properties of electrons and holes.

This is because the superlattice has a modified dispersion relation in the conduction band as well as in the heavy-hole valence band, the light-hole valence band, and the spin-orbit split-off valence band. The thermionic field emission and tunneling transport of electrons and holes from the superlattice layer to the barrier layer also determine the JV characteristics of these solar cells. This is also true in the case of MQW solar cells. Figures 8.3, 8.5, and 8.6 show some of the multi-quantum well and superlattice-based InGaN solar cells in the literature. Figure 8.4 shows typical characteristic of a graded composition InGaN solar cell.

Complex Dielectric Function

The complex dielectric function determines the photon velocities in the InGaN material and the absorption coefficient, as discussed in Chapter 7. The imaginary part of the dielectric function, which determines the absorption coefficient, is typically very low below the bandgap and acquires large, positive values above the bandgap. Complex dielectric functions can be calculated from the band structures, and they can be experimentally measured using ellipsometric measurements.

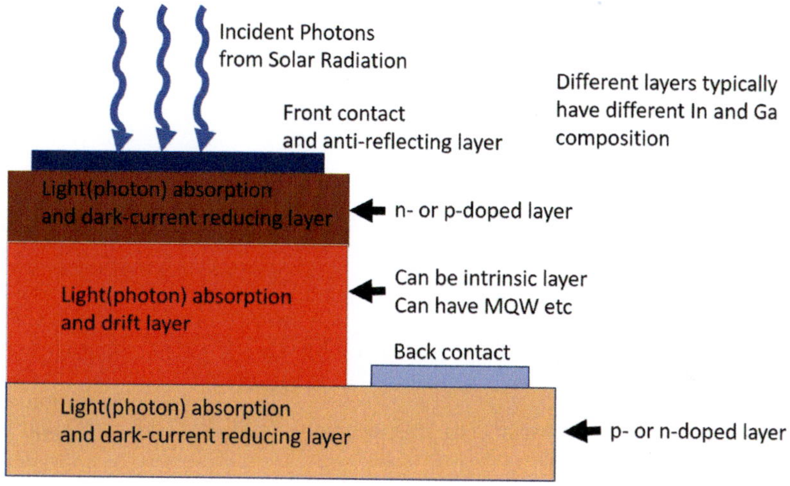

Figure 8.3 The typical structure of an InGaN photovoltaic cell.

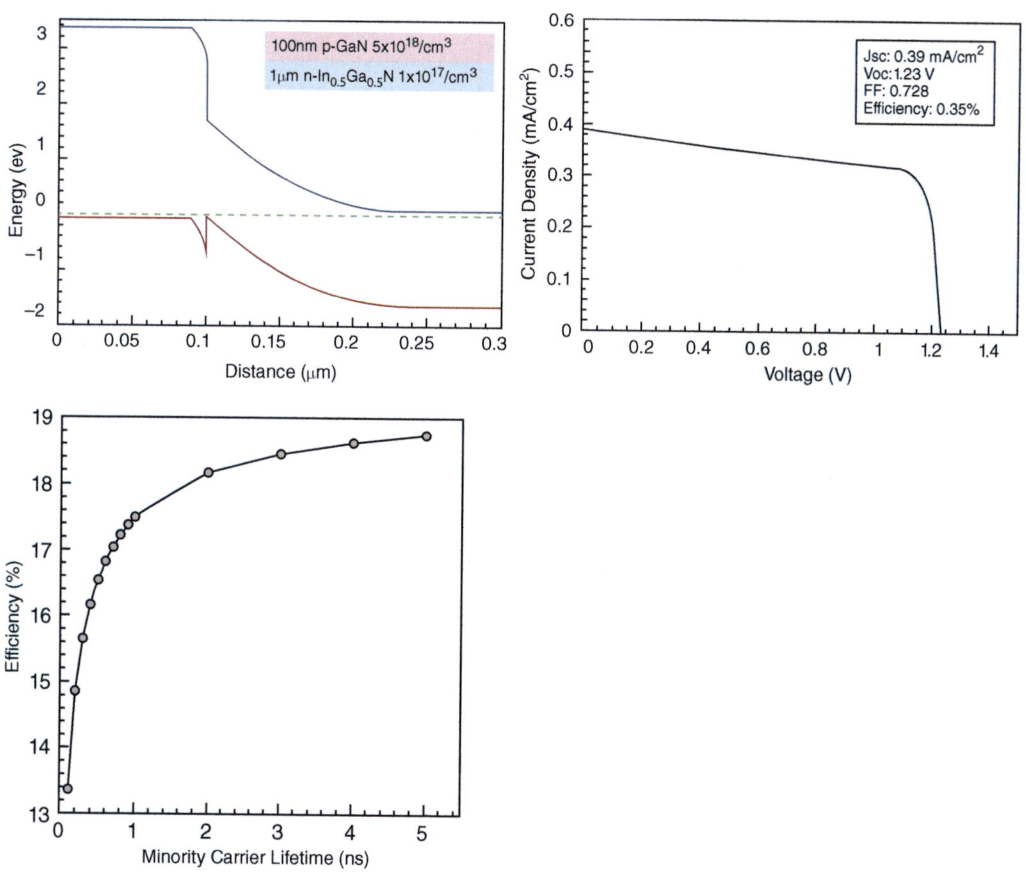

Figure 8.4 *JV* characteristics and PCEs of compositionally graded InGaN solar cells, after Reference [3], used with permission.

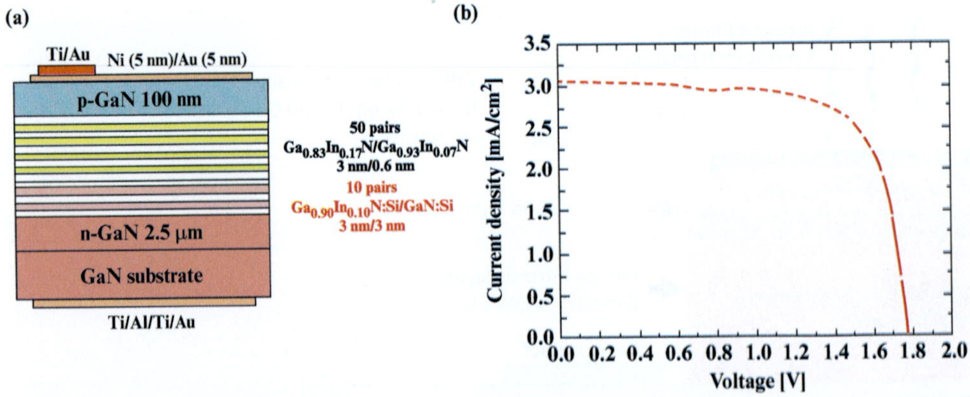

Figure 8.5 *JV* characteristics of superlattice InGaN solar cells, after References [4, 5], used with permission.

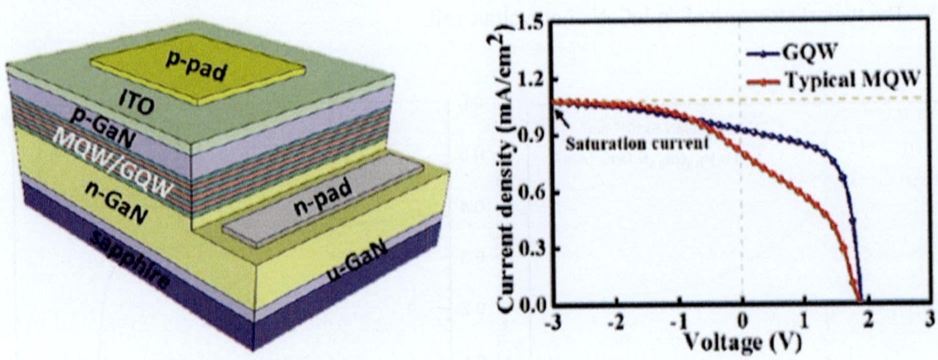

Figure 8.6 *JV* characteristics of MQW-based InGaN solar cells, after Reference [6], used with permission.

Many InGaN solar cell designs use composition grading to obtain the best efficiencies. To the first order, material composition grading can be expected to result in corresponding grading of the complex dielectric function. However, if the grading is too aggressive, non-local effects on the complex dielectric function can result in non-trivial changes to the absorption profile in any solar cell device. This necessitates more careful atomistic calculation of complex dielectric functions and absorption spectra in these devices.

Atomistic effects on complex dielectric function can become even more complicated for intermediate band solar cells (IBSCs), discussed later in the section on IBSCs. These solar cells have quantum dot (QD) layers or quantum-well layers, and the absorption spectrum gets modified significantly compared to bulk layers. QD layers have very discrete absorption spectra and quantum wells also show the effect of discreteness due to significant quantization. Time-dependent perturbation theory of quantum mechanics can be used to obtain the complex dielectric functions in these cases.

Polarization Grading in *p-n* Junction Solar Cells

Composition grading in *p-n* junction group III-Nitride solar cells leads to polarization grading. This is because the sum of spontaneous and piezoelectric polarization is a function of indium concentration in InGaN. While designing composition graded solar cells using group III nitrides, this polarization grading must be kept in minds. Polarization grading leads to bound charge densities and it affects the band bending through the Poisson equation. Advanced finite-element simulations can become necessary to take these polarization effects into account in solar cell design.

If polarization effects become detrimental for some solar cells grown along the *c*-axis, *m*-plane, and *a*-plane growth can lead to designs where there is no polarization along the growth axis, which is also the major direction of current flow. However, *m*-plane and *a*-plane growth is more difficult to optimize, and these kinds of devices come with higher dislocation densities and higher stacking fault densities. These dislocations and stacking faults lead to non-radiative recombination and reduced quantum efficiencies for these solar cells.

Polarization grading in alloys like InGaN can lead to significant polarization fluctuations at the microscopic or mesoscopic scale. This can lead to extra scattering from the polarization-Coulomb fluctuations, which leads to the degradation of mobility and a decrease in diffusion coefficients. From the solar cell physics transport equations discussed previously, this kind of extra scattering tends to decrease device efficiency. So, this effect must be quantified carefully to understand if extra Coulombic scattering dominates over the positive aspects of polarization grading or whether the positive aspects dominate.

Schottky Barrier Solar Cells

Schottky barrier solar cells can help achieve some design simplifications and process optimizations using a unipolar design. Composition grading can be used to obtain optimized power absorption characteristics and good power-conversion efficiencies (References [7, 8]). The Schottky barrier height to transparent conducting electrode and polarization grading (spontaneous and piezoelectric polarization) play a key role in these designs. Depletion width needs to be optimized alongside position-dependent absorption spectrum to obtain best characteristics.

Transparent conducting electrodes (TCEs) are needed to obtain good power-conversion efficiencies using Schottky barrier solar cells. Indium tin oxide (ITO) is the preferred TCE because of its low absorption in the solar spectrum. TCEs have plasma edges that are in the far infrared or mid-infrared, so that they transmit most of the light in the solar spectrum.

One advantage of Schottky barrier solar cells is that it can be designed with *n*-type layers only and factors like low magnesium (Mg) *p*-type dopant activation do not affect device characteristics. Using composition grading in Schottky barrier solar cells can help decrease hot carrier energy loss to phonons. The indium composition is lowest at the end where the solar photons are incident and the indium composition is increased away from the front interface. High energy solar radiation is absorbed nearer to the front interface and lower energy radiation is absorbed away from the front interface, and the electric field in the depletion layer helps sweep the photogenerated electrons and holes away to be collected at the electrodes.

The ratio of photogenerated current to the dark current of the Schottky barrier diode is the crucial factor effecting cell efficiency. Therefore, efforts must be made not only to increase the photogenerated current but also to reduce the dark current. The Schottky barrier height becomes crucial to achieving low dark current. The right kind of transparent conducting electrode with large Schottky barrier height becomes necessary to achieve good efficiencies.

Intermediate Band Solar Cells

Typical photovoltaic cells based on *p*-*n* or *p*-*i*-*n* junctions need careful optimization between several device parameters like recombination rates of photogenerated carriers, *p*-layer width, *n*-layer width, depletion width, etc. The solar spectrum photons must be all absorbed within the device and the photogenerated carriers must be collected before recombining. Further, secondary photons emitted during band-to-band recombination must be reabsorbed to obtain good efficiencies approaching the Shockley–Queiser limit. It can become beneficial to increase the absorption coefficients in the infrared and red part of the solar spectrum. IBSCs use quantum structures like QDs and quantum wells to obtain higher infrared photon absorption.

When QD layers or quantum-well layers are used, there can be second photon absorption, which leads to higher absorption coefficients. Second photon absorption refers to the phenomenon when the electron generated in the conduction band of the QD absorbs a second photon to get excited into continuum states of the QD. This is possible because the $\boldsymbol{k}$-vector conservation constraint is absent in a quantum-confined structure like a QD. The same argument holds for quantum-well layers also. This second photon absorption is typically helpful to increase the absorption coefficient for the infrared and red end of the solar spectrum. The blue end and the UV parts are not affected much by the IBSC design, and these components have the usual absorption coefficients. However, the effect of this change in absorption coefficient for some parts of the solar spectrum must be quantified carefully to understand the advantages for any specific design. The need for multiple QD layers or multiple quantum-well layers depends on the exact way in which absorption coefficients are modified in the infrared (IR) and red ends of the spectrum.

In group III-Nitride IBSCs based on QDs, QD layers with high In content are grown in specific layers and they are cladded by lower In-content layers. The growth of QDs can be very sensitive to the combination of In content and required dimensions while using the Stranski–Krastanov mode. Growing the right type of high In-content QDs can be quite a challenge. Further, molecular beam epitaxy growth of QD layers adds process complexity and cost. It may become preferable to obtain chemical vapor deposition QD layers, but they are much more difficult to obtain.

InGaN QDs tend to have less serious built-in field effects from polarization discontinuities compared to quantum wells. However, the built-in fields can still significantly alter the absorption characteristics. Therefore, QD-based IBSCs require as much device optimization as quantum well-based IBSCs.

One constraint in QD IBSCs is that QDs have very discrete energy levels below the continuum states. Therefore, the first photon absorption takes place in very discrete ranges. That means a large part of the solar radiation cannot be used for the second photon absorption-based improvement to PCE. QD dimension optimization can help address this problem to some extent. However, this is still a subject of research and the right balance between QD growth conditions and device requirements must be achieved.

The fabrication of group III-Nitride IBSCs based on quantum wells also needs careful optimization between device requirements and growth constraints. High In-content quantum wells have very large built-in electric fields and a large quantum-confined Stark effect due to polarization. Therefore, the indium content in the well layer and the barrier layer must be chosen carefully to obtain the right absorption characteristics. A $2\times10^6\,\mathrm{V\,cm^{-1}}$ built-in field in a 5 nm quantum well leads to a 1 eV Stark shift. Efforts must be made to account for this Stark shift while designing IBSC photovoltaic devices. Another constraint in quantum-well IBSCs is the large lattice mismatch between typical substrates and high In-content layers. These lattice mismatches result in high dislocation densities, which lead to deep states for non-radiative recombination. Non-radiative recombination has a very adverse impact on solar cell power efficiencies since carriers are lost to sub-bandgap photons that cannot be recovered.

The use of superlattice-type template growth can help reduce dislocation densities but they add significant process complexity. The thickness of the quantum-well layers and the thickness of the barrier layers are also design optimization parameters because the amount of piezoelectric polarization depends on the amount of strain relaxation in these layers. Completely pseudomorphic layers have the maximum amount of piezoelectric polarization while layers with strain relaxation have lower piezoelectric polarization.

Another potential design dimension for IBSCs is the growth direction. If *c*-plane growth leads to QDs with polarization effects that prove detrimental to solar cell efficiency, *m*-plane and *a*-plane QDs can be explored as alternatives. However, even *m*-plane and *a*-plane QDs also have polarization effects so we are looking not at complete elimination of polarization effects but modification of polarization effects.

The polarization-Coulomb field due to polarization fluctuations due to alloy disorder can lead to extra microscopic Coulombic scattering near the QD-barrier interface. The effect of this needs to be quantified to understand if the QD layers lead to higher overall efficiencies. The same kind of polarization-Coulomb fluctuations are also relevant in designs with multiple quantum-well layers. Various types of heterojunction solar cells and IBSCs are discussed in References [9–17] and References [18–22].

Substrate Effects on InGaN Solar Cells

The substrate has a large effect on device quality because it decides the dislocation densities in the device layers and the resulting non-radiative recombination. It also has a significant effect on the cost of the solar cells. For example, SiC substrates have better lattice matching with the nitride semiconductors compared to sapphire. However, they are much more expensive.

Further, there is a lot of interest in obtaining group III-Nitride devices in Si substrates because of cost benefits. However, Si also has a large lattice mismatch to the group III nitrides, and therefore, cost and device quality considerations must be balanced. Since polarization along the *c*-axis complicates device design for typical *c*-plane growth, there is interest in non-polar axes growth and growing cubic nitride layers. However, non-polar *a*-plane and *m*-plane growth is more expensive and technologically challenging compared to *c*-plane growth. On Si substrates, growth on Si (111) face typically leads to *c*-plane wurtzite growth of nitrides while growth on Si (100) face leads to cubic nitride semiconductor growth. However, the wurtzite crystal structure is much more stable for group III nitrides compared to the cubic structure. Therefore, when cubic InGaN is grown on Si (100) substrate, significant domain formation with wurtzite and cubic domains can become a challenge.

The use of dislocation reduction layers in template designs is of potential interest in the design of solar cells. Template designs basically use intermediate layers that are better matched to the substrate. This can lead to substantial reduction of dislocation densities and stacking fault densities in the device layers.

The reduction of dislocation densities using the right kind of substrate engineering is necessary to reduce non-radiative SRH-type recombination. Non-radiative recombination leads to a loss of electrical energy in the solar cell and should be avoided at all costs.

Another aspect of substrate engineering that is relevant to solar cell design is the different amounts of defect densities for *c*-plane growth as opposed to *m*-plane or *a*-plane growth. Typically, growth technology for *c*-plane growth is highly advanced so that *a*-plane and *m*-plane growth have a higher density of dislocation densities and stacking fault densities. Therefore, any design that uses *a*-plane or *m*-plane growth to get better polarization engineering must consider a higher non-radiative recombination from dislocations and stacking faults. The design should be preferred only if the effects of higher non-radiative recombination are outweighed by the polarization engineering aspects. Substrate effects on group III-Nitride solar cells can be found in References [23–35].

Ohmic Contact Effects in *p-n* and *p-i-n* InGaN Solar Cells

The *p*-type layer in *p-n* and *p-i-n* InGaN solar cells offers technological difficulties due to two major reasons. One is that the Mg *p*-type dopant has a low ionization ratio in InGaN when the In content is low (low In-content layers can become necessary to get good efficiency based on the device equations). This means that *p*-type layers tend to have a higher series resistance compared to other material systems. This can decrease power efficiency substantially.

Also, *p*-type ohmic contacts in InGaN tend to have high specific contact resistance because hole masses are larger compared to other semiconductor systems and tunneling transmission probabilities at the ohmic contacts are low. Using grading to obtain lower ohmic contact resistance adds significant process complexity.

Plasmonically Enhanced Solar Cells

Using metallic nanostructures can help achieve plasmonic enhancement of absorption and altered solar cell current-voltage characteristics. The transport of carriers from metallic nanostructures into the semiconductor and into the collector metal electrodes becomes crucial to determining if this kind of design leads to any overall advantage over conventional designs that do not use plasmonic nanostructures.

The absorption spectrum of the silver (Ag) nanodisk-based plasmonic solar cell in Figure 8.7 depends on the surface plasmon resonance (SPR) excitation in the nanodisk and the energy relaxation mechanisms in the nanodisk. In principle, it is possible to tune the SPR spectrum and the absorption spectrum by varying nanodisk dimension for small nanodisk dimensions. Using this design parameter, one can design the plasmonic solar cell to have good absorption efficiency in specific parts of the solar spectrum. Typically, the SPR resonance in these nanodisks are in the infrared region. Therefore, the plasmonic nanodisk tends to absorb the IR part of the solar spectrum while the visible and UV parts are absorbed in the *p-n* junction part. Figure 8.8 shows typical characteristics of plasmonically enhanced InGaN solar cells.

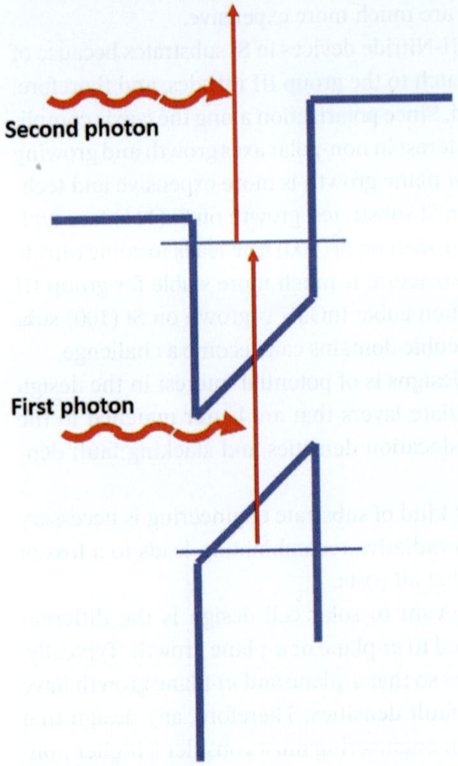

Figure 8.7 First photon and second photon absorption in IBSCs.

Solar Concentrating Photovoltaics

Solar concentrating photovoltaic systems use a combination of solar power concentration and semiconductor photovoltaic devices to reach efficiencies as high as 30% and potentially higher.

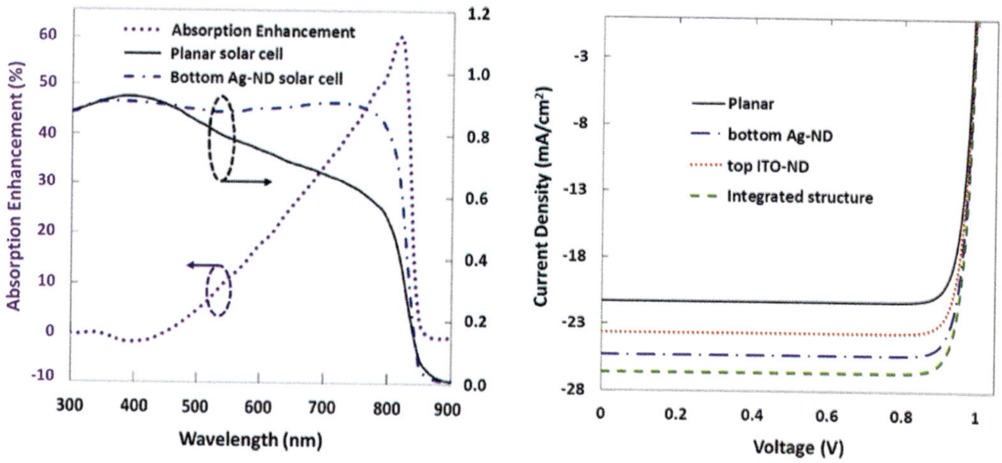

Figure 8.8 Absorption enhancement and *JV* characteristics of plasmonically enhanced InGaN solar cells, after Reference [36], used with permission.

Reference [37] demonstrates efficiencies of about 1.5% from room temperature up to 700 K using InGaN/GaN multi quantum-well structure and solar power concentration. However, there is more scope for optimization with these concentrated power photovoltaic schemes. For example, other kinds of composition grading and QD schemes can potentially achieve higher efficiencies by better control over quantum efficiencies and device parameters like dark current that affect open-circuit voltages and eventually power-conversion efficiencies.

The device equations in the solar cells have various competing terms like optical generation, drift and diffusion, and radiative and non-radiative recombination. Solar concentration offers an additional design handle by changing the optically generated carriers part of the device physics.

However, higher generation rates also go with higher recombination and one needs to calculate these terms and their relative magnitudes carefully to reach the best design. Figure 8.9 shows typical concentrator photovoltaic characteristics.

Tandem Solar Cells Using Indium Gallium Nitride

Since the InGaN alloy system covers the spectral range from the infrared through the visible to the UV, it is possible to make tandem solar cells with three or four modules where each module absorbs solar radiation in a specific narrow wavelength range and converts it efficiently into electrical power. Reference [38] reports simulation results of tandem solar cells with efficiencies around 3%. However, further optimization should be able to obtain efficiencies much higher than the Shockley–Queiser limits of single-junction solar cells.

Tandem cells obtain higher efficiencies compared to the single-junction Shockley–Quieser limit because they use different sub-cells to absorb specific regions of the solar spectrum, as shown in Figure 8.10. This leads to a situation where most of the photogenerated carriers are close to the conduction-band minimum or valence-band maximum. This means hot carrier energy loss to phonons (for absorption much above the bandgap energy) can be minimized in tandem cell designs. However, this advantage in terms of minimizing hot carrier energy losses comes at the cost of process complexity.

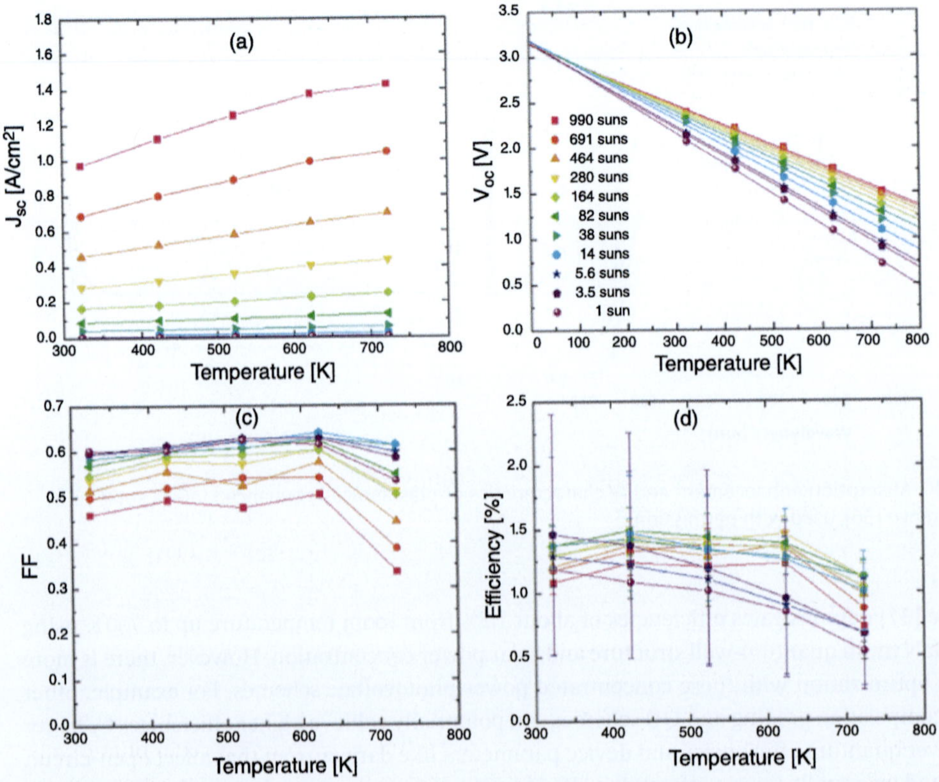

Figure 8.9 Short-circuit currents, open-circuit voltages, fill factors, and power-conversion efficiencies of multi quantum-well concentrator photovoltaic scheme. After Reference [37], used with permission.

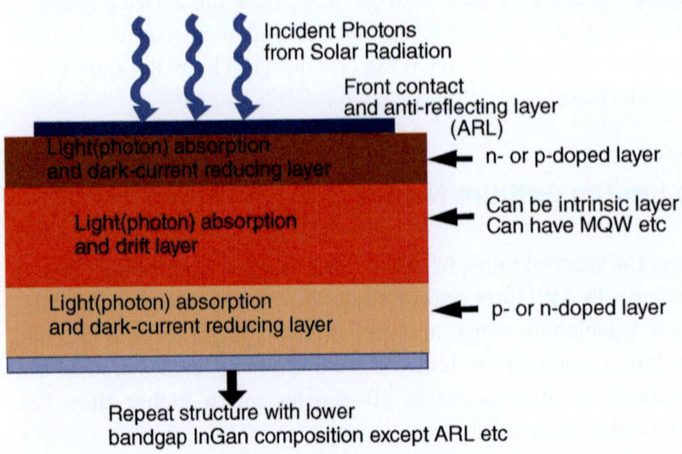

Figure 8.10 Typical tandem cell configuration.

A major constraint in obtaining tandem InGaN solar cells with good efficiency is the contact between various cells in the solar cell. Typically, in a tandem cell the contact between the different sub-cells depends on tunnel junctions between the cells. The tunnel junctions realize large tunneling probabilities of carriers in p++/n++ depletion layers. However, the nitride semiconductor system tends to have large carrier masses compared to other common semiconductors.

Large carrier masses lead to the exponential suppression of tunneling current, and therefore, good tunnel barriers are difficult to obtain. Additional material grading is needed to enable high-efficiency tandem solar cells. However, using additional material grading just to obtain good tunneling barriers between the sub-cells is a significant addition to process and device design complexity. Also, superlattice schemes can potentially lead to better tunnel junctions for these tandem solar cells.

Typical tandem cells in production tend to use different material system to focus on absorbing specific parts of the solar spectrum. This adds to process and design complexity. The InGaN material system clearly has an advantage in this regard because it spans the entire solar spectrum range. However, factors like the low activation of Mg *p*-type dopants add process complexity. Superlattice tricks can sometimes lead to higher activation of Mg *p*-type dopant activation and to overall better performance of the tandem cells. Tandem cell designs are typically preferred when efficiencies are more important than cost and process complexity.

When three or four sub-cells are used in the tandem cell with one subcell being made of InGaN with high Ga content, this specific subcell also has the problem of high ohmic contact resistance. Getting good ohmic contacts to this subcell can involve using thin cap layers with polarization-induced built-in fields for enhanced tunneling. The other option is to use graded layers near the ohmic contact. Either way, significant process complexity needs to be contended with to achieve InGaN tandem solar cells.

Another question that needs to be considered in InGaN tandem solar cell design is whether *c*-plane growth is to be preferred or whether *a*-plane and *m*-plane growths cam be preferred. Since there are so many complex factors affecting solar cell efficiency, the answer to this question needs further research. Using superlattice tricks to obtain better tunnel junctions and better ohmic contacts is easier for *c*-plane growth, but *a*-plane growth and *m*-plane growth may offer better options for optimizing the carrier transport layers of the solar cell.

Semiconductor Photocatalysis Using InGaN

Indium gallium nitride QDs can be used as photocatalytic material to perform photocatalytic electrolysis of water (H_2O) to generate hydrogen (H_2) and oxygen (O_2). This is of potential use in fuel cell vehicles and fuel cell energy generation as a clean fuel alternative. Heterojunction variations can be used in these QDs to enhance charge separation characteristics so that the photogenerated electrons and holes can more efficiently generate H_2 and O_2 by oxidation and reduction processes. The fact that the bandgap of the InGaN alloy system spans the range from infrared to UV through the visible means that different composition QDs can be used to optimize absorption and conversion to H_2 and O_2. The nature and distribution of surface states in QDs play a crucial role in determining the efficiency of the photocatalysis scheme because they play a crucial role in how photogenerated holes can drift/diffuse to the surface and act as oxidizing or reducing agents to generate H_2 and O_2.

Also, InGaN QDs have some amount of polarization-induced quantum-confined Stark effect (QCSE). This can lead to efficient separation of electrons and holes as well as better conversion efficiencies. The QDs can be realized as dot-in-wire versions where the dots exist inside nanowires. This kind of design is useful because nanowires can be grown using some simplified growth processes.

Directly using nanowires as photocatalytic components is also possible with proper design. One way is to use nanowires grown perpendicular to the *c*-axis so that the polarization is perpendicular to the nanowire axis. The photogenerated electrons and holes can then be separated perpendicular to the axis and that can cause electrolysis of water. Nanowire density becomes a major parameter when we use this method to achieve photocatalytic electrolysis of water.

The use of dislocation-free QDs and nanowires becomes necessary for good photocatalytic activity. This is because dislocations lead to mid-gap recombination centers, which lead to non-radiative recombination of electrons and holes. Since we want photogenerated electrons and holes to cause electrolysis of water rather than recombine, we need to keep these dislocation densities to a minimum. Therefore, growth optimization becomes necessary to get relatively dislocation-free QDs and quantum wires.

References

1 Luque, A. and Hegedus, S. (eds.) (2011). *Handbook of Photovoltaic Science and Engineering*. J.Wiley and Sons.

2 Kazazis, S.A. et al. (2017). Polarization-engineered InGaN/GaN solar cells: realistic expectations for single heterojunctions. *IEEE J. Photovolt.* 8 (1).

3 Brown, G.F. et al. (2010). Finite element simulations of compositionally graded InGaN solar cells. *Sol. Energy Mat. Sol. Cells* 94.

4 Bhuiyan, A.G. et al. (2012). InGaN solar cells: present state of the art and important challenges. *IEEE J. Photovolt.* 2 (3).

5 Kuwahara, Y. et al. (2011). GaInN based solar cells using strained layer GaInN/GaInN suoerlattive active layer on a freestanding GaN substrate. *Appl. Phys. Exp.* 4 (2).

6 Tsai, Y. et al. (2015). Enhanced power conversion efficiency in InGaN-based solar cells via graded composition multiple quantum wells. *Opt. Exp.* 23 (24).

7 Praharaj, C.J. (2007). Optical absorption at digitally and continuously graded Indium Gallium Nitride Schottky barriers. *Mater. Res. Soc. Proceed.* 1012.

8 Praharaj, C.J. (2021). Indium Gallium Nitride on silicon heterojunction Schottky barrier solar cell characteristics. *MRS Commun.* 11.

9 Seo, T.H., Shim, J.P., Chae, S.J. et al. (2013). Improved photovoltaic effects in InGaN-based multiple quantum well solar cell with graphene on indium tin oxide nanodot nodes for transparent and current spreading electrode. *Appl. Phys. Lett.* 102.

10 Dahal, R., Li, J., Aryal, K. et al. (2010). InGaN/GaN multiple quantum well concentrator solar cells. *Appl. Phys. Lett.* 97.

11 Lee, H.C., Su, Y.K., Lan, W.H. et al. (2011). Study of electrical characteristics of GaNbased photovoltaics with graded Inx Ga1−xN absorption layer. *IEEE Photon. Tech. Lett.* 23 (6).

12 Fujii, T., Kuwahara, Y., Iida, D. et al. (2011). GaInN-based solar cells using GaInN/GaInN superlattices. *Phys. Stat. Sol. (C)* 8.

13 Liou, B.W. (2011). Design and fabrication of Inx Ga1 −x N/GaN solar cells with a multiple-quantum well structure on SiCN/Si(111) substrates. *Thin Solid Films* 520 (3).

14 Kuwahara, Y., Fujii, T., Sugiyama, T. et al. (2011). GaInN-based solar cells using strained-layer GaInN/GaInN superlattice active layer on a freestanding GaN substrate. *Appl. Phys. Exp.* 4.

15 Braun, A., Katz, E.A., and Gordon, J.M. (2013). Basic aspects of the temperature coefficients of concentrator solar cell performance parameters. *Prog. Photovolt. Res. Appl.* 21 (5).

16 Thosar, M. and Mathew, M. (2013). Modeling for high efficiency Gan/Ingan solar cell. *IOSR J. Electr. Electron. Eng.* 4 (1).

17 Lai, K.Y., Lin, G.J., Lai, Y.-L. et al. (2010). Effect of indium fluctuation on the photovoltaic characteristics of InGaN/GaN multiple quantum well solar cells. *Appl. Phys. Lett.* 96.

18 Neufeld, C.J., Cruz, S.C., Farrell, R.M. et al. (2011). Effect of doping and polarization on carrier collection in InGaN quantum well solar cells. *Appl. Phys. Lett.* 98.

19 Fu, P.H., Lin, G.J., Ho, C.H. et al. (2012). Efficiency enhancement of InGaN multi-quantum-well solar cells via light-harvesting SiO2 nano-honeycombs. *Appl. Phys. Lett.* 100.

20 Chouchen, B. et al. (2019). Numerical modeling of the electronic and electrical characteristics of InGaN/GaN-MQW solar cells. *MDPI Mater.* 12 (8).

21 Lee, Y.-J., Lee, M.H., Cheng, C.-M., and Yang, C.-H. (2011). Enhanced conversion efficiency of InGaN multiple quantum well solar cells grown on a patterned sapphire substrate. *Appl. Phys. Lett.* 98.

22 Jeng, M.-J., Lee, Y.-L., and Chang, L.-B. (2009). Temperature dependences of Inx Ga1−xN multiple quantum well solar cells. *J. Phys. D, Appl. Phys.* 42.

23 Wierer, J.J., Jr., Fischer, A.J., and Koleske, D.D. (2010). The impact of piezoelectric polarization and nonradiative recombination on the performance of (0001) face GaN/InGaN photovoltaic devices. *Appl. Phys. Lett.* 96.

24 Wierer, J.J. et al. (2010). The impact of piezoelectric polarization and nonradiative recombination on the performance of (0001) face GaN/InGaN photovoltaic devices. *Appl. Phys. Lett.* 96.

25 Matioli, E., Neufeld, C., Iza, M. et al. (2011). Speck, and C. Weisbuch, High internal and external quantum efficiency In- GaN/GaN solar cells. *Appl. Phys. Lett.* 98.

26 Bi, Z., Bacon-Brown, D., Du, F. et al. (2018). An InGaN/GaN MQWs solar cell improved by a surficial GaN nanostructure as light traps. *IEEE Photonics Technol. Lett.* 30 (1).

27 Shim, J.-P., Choe, M., Jeon, S.-R. et al. (2011). InGaNbased p–i–n solar cells with graphene electrodes. *Appl. Phys. Exp.* 4 (5).

28 Young, N.G., Perl, E.E., Farrell, R.M. et al. (2014). High-performance broadband optical coatings on InGaN / GaN solar cells for multijunction device integration. *Appl. Phys. Lett.* 104.

29 Kuwahara, Y., Fujii, T., Fujiyama, Y. et al. (2010). Realization of nitride-based solar cell on freestanding GaN substrate. *Appl. Phys. Exp.* 3 (11).

30 Ali, N. et al. (2020). Advances in nanostructure homojunction solar cells and photovoltaic materials. *Mater. Sci. Semiconductor Process* 107.

31 Braun, A., Hirsch, B., Vossier, A. et al. (2013). Temperature dynamics of multijunction concentrator solar cells up to ultra-high irradiance. *Prog. Photovolt. Res. Appl.* 21 (2).

32 Redaelli, L. et al. (2015). Effect of the barrier thickness on the performance of multiple-quantum-well InGaN photovoltaic cells. *Jpn. J. Appl. Phys.* 54: 072302.

33 Yang, C.C., Sheu, J.K., Liang, X.-W. et al. (2010). Enhancement of the conversion efficiency of GaN-based photovoltaic devices with Al-GaN/InGaN absorption layers. *Appl. Phys. Lett.* 97.

34 Lestrade, M., Li, Z., Xiao, Y., and Li, Z.M.S. (2011). Modeling of polarization effects in ingan pin solar cells. *Opt. Quant. Electron* 42.

35 Sang, L. et al. (2015). InGaN-based thin film solar cells: epitaxy, structural design and photovoltaic properties. *J. Appl. Phys.* 117.

36 Kumawat, U.K. et al. (2020). Plasmonic and Photonic Enhancement of photovoltaic characteristics of indium-rich InGaN p-n junction solar cells. *Opt. Exp.* 28 (8).

37 Moses, G. et al. (2020). InGaN/GaN multi-quantum-well solar cells under high solar concentration and elevated temperatures for hybrid solar thermal-photovoltaic power plants. *Prog. Photovolt.* 28 (11).

38 Boudaoud, C. et al. (2020). Simulation and optimization of a tandem solar cell based on InGaN. *Math. Comp. Simul.* 167.

9

Quantum Photonic Properties of Nitride Semiconductor Devices

Non-classical Light from Group III-Nitride Heterostructures

Obtaining non-classical photonic emission or light from semiconductor nanostructures has become a subject of intense research interest because of their potential use in quantum computation. Semiconductor quantum dots (QDs) can be used to obtain single-photon emissions, one of the ways being to excite low intensity light to create a single exciton that recombines to emit a single photon. Group III-Nitride QDs are of great interest because they can provide single-photon sources from the infrared to the ultraviolet energy range through the visible. Various quantum-computing frameworks can require controllable photon emission in different energy ranges for a variety of reasons like noise immunity and error rate control. By tuning alloy composition in group III-Nitride QDs, the energy range of emitted photons can be tuned from 0.7 eV to 6.2 eV. The spectral purity of the photon can also be tuned by engineering the natural linewidth and inhomogeneous broadening of the photon emission. Therefore, tuneable quantum emission and detection become integrable in a single material system using the group III-Nitride semiconductors.

Group III-Nitride Single-photon Quantum Emitters

Indium gallium nitride (InGaN) QDs have been used to obtain single-photon emitters in a range of wavelengths. These emitters have use in quantum computation and potential use as quantum memories. The second-order intensity autocorrelation is used to detect single-photon emission because this function $g^{(2)}(\tau)$ shows a prominent dip at $\tau = 0$ when actual single-photon emission is achieved.

$$g^2(\tau) = \frac{< I(t) I(t+\tau) >}{< I(t) >^2} \tag{9.1}$$

Reference [1] uses coincidence counts that translate to an equivalent of $g^{(2)}(0)$ being close to zero (0.043) and indicative of single-photon emission from an InGaN QD (Figures 9.1 and 9.2). There is a large scope for optimization by varying dot dimensions and polarization magnitudes when it comes to the energy of the photon and the emission lifetime. Molecular beam epitaxy (MBE) growth in the Stranski–Krastanow mode is preferred for obtaining InGaN QDs. However, a lot of optimization is needed to obtain desired wavelengths (and equivalent energies) and uniformity. The inhomogeneous broadening due to dot-dimension fluctuations needs careful attention in terms of optimizing growth uniformity.

Group III-Nitride Semiconductor Optoelectronics, First Edition. C. Jayant Praharaj.

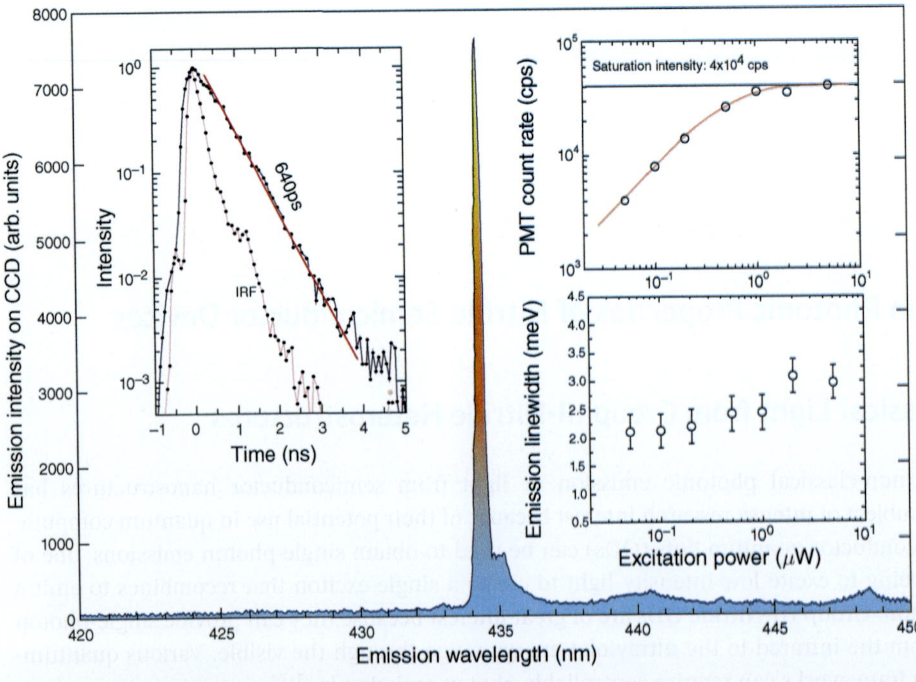

Figure 9.1 The emission spectrum of InGaN QD at an excitation power of 0.2 μW. Emission lifetime indicates a small QD. After Reference [1].

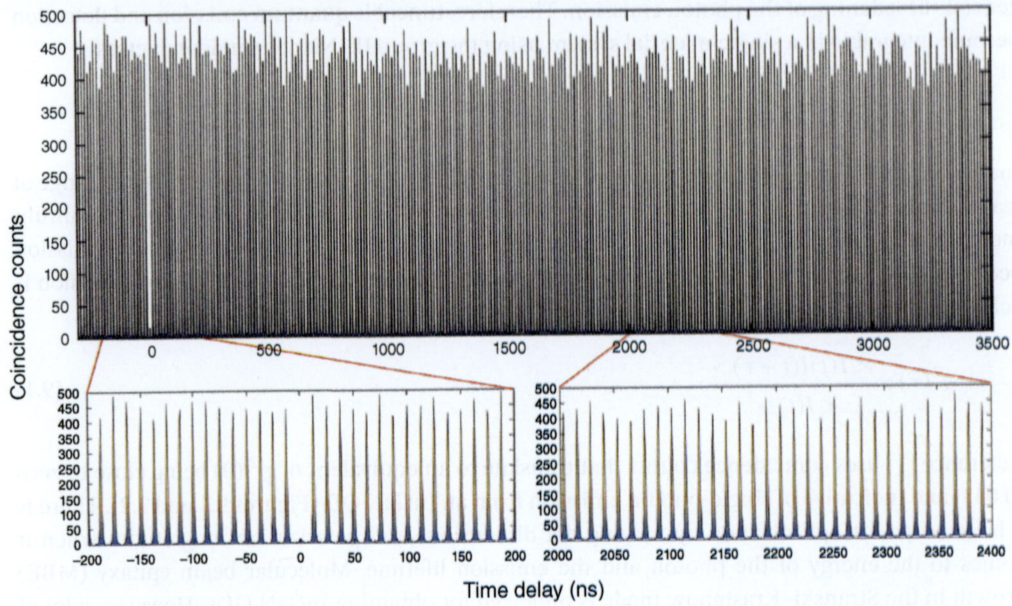

Figure 9.2 Coincidence counts as a function of time delay indicating anti-bunching and single-photon emission. After Reference [1].

Excitons in Group III-Nitride Quantum Dots

Since InGaN QDs have been used to obtain single-photon emitters there is a lot of interest in determining the energy ranges of obtainable photons both theoretically and experimentally. The confinement of electrons and holes in the QD typically gives rise to a large Coulomb attraction energy between electron and hole that forms the exciton. However, as discussed in Chapter 4, the built-in electric fields due to polarization discontinuities in InGaN heterostructures lead to decreased overlap between electron and hole wave functions. This effect leads to decreased Coulomb interaction energy in InGaN QDs compared to QDs of comparable dimensions in other material systems. Therefore, the bandgap renormalization magnitude is decreased. It must be kept in mind that these polarization-induced fields exist in wurtzite InGaN/gallium nitride (GaN) and wurtzite GaN/AlGaN QDs. The wurtzite structure is the most common technologically and the most thermodynamically stable. The cubic version of these QDs is much harder to fabricate, and therefore, the polarization-induced Coulomb energy decrease should be considered in most practical InGaN or GaN QDs.

The direct Coulomb interaction energy between the conduction-band electron and valence-band hole is given by

$$E_{direct} = \iint \left|\psi_c(\boldsymbol{r}_1)\right|^2 \left|\psi_v(\boldsymbol{r}_2)\right|^2 \frac{e^2}{4\pi\varepsilon\left|\boldsymbol{r}_1 - \boldsymbol{r}_2\right|} d^3\boldsymbol{r}_1 d^3\boldsymbol{r}_2 \tag{9.2}$$

There is also an exchange interaction between the electrons and holes. This is more complex because some of it is subsumed in the Bloch eigenstates in the crystals that form the QD. The reader is encouraged to read the literature on the subject to gain a better understanding of the excitonic binding energy in InGaN and GaN QDs. Figure 9.3 shows excitonic emission from a GaN QD with fine structure splittings. Figures 9.4 and 9.5 show biexcitonic and temperature effects, respectively, in group III-Nitride QDs. The reader can find a wide range of QD emission characteristics in References [3–25]. Temperature-dependent quantum emitters are discussed in References [26–33].

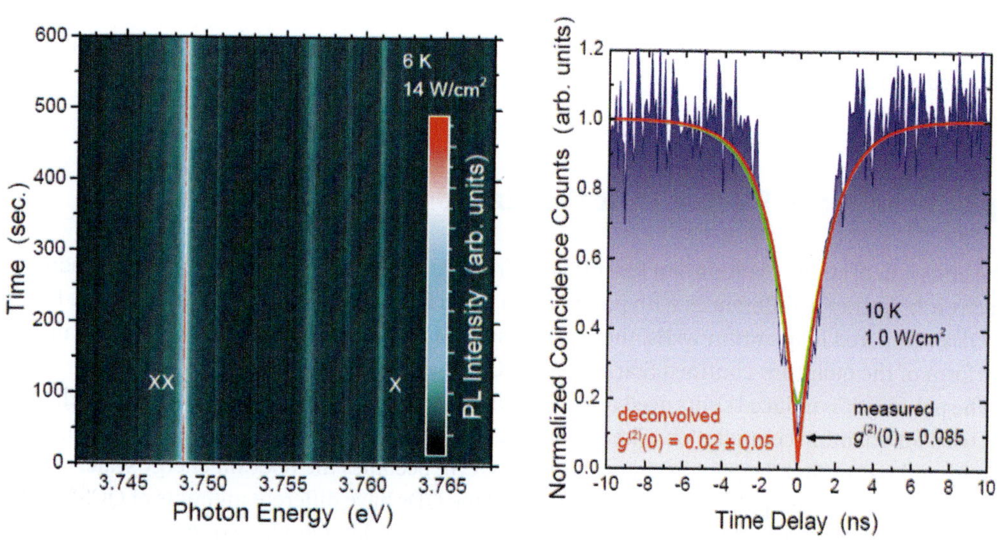

Figure 9.3 Exciton emission with fine structure splitting in GaN QD (after Reference [2]) and second-order intensity autocorrelation.

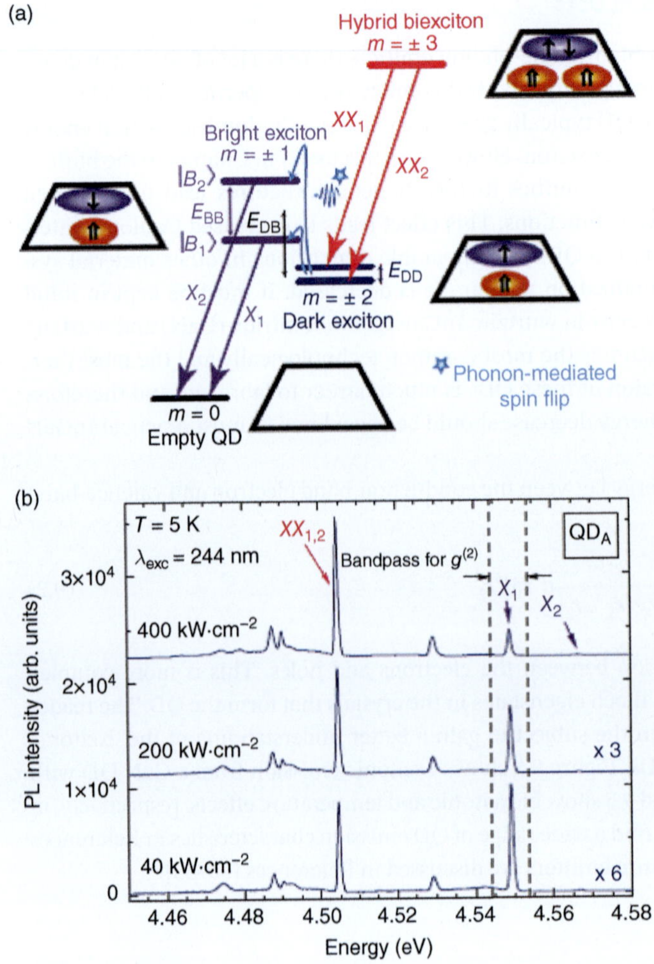

Figure 9.4 Biexcitonic double-photon emission from GaN/aluminum nitride (AlN) QDs. From Reference [34], used with permission.

Spontaneous and Piezoelectric Polarization Effects

The emission of single photons or a few photons from QDs happens at energies that correspond to electron-hole energy differences with polarization effects. While the piezoelectric effects are lower in QDs compared to quantum wells, there is still substantial polarization-induced energy shifts in the form of the quantum-confined Stark effect (QCSE).

The polarization-induced shifts need to be accounted for when the photons produced from the QD are used for quantum computation. The quantum-computing system imposes its own constraints on energy and purity, and the photon energy with the QCSE becomes the relevant energy scale.

c-Plane QDs and non-polar *a*-plane QDs and *m*-plane QDs have different amounts of QCSE shifts in energy. It is possible to find some expensive substrates on which *m*-plane or *a*-plane InGaN can be grown without strain. It has the potential to engineer the photon emission according to system requirements. Strain relaxation by use of buffer layers before growing the device layers (QD layers) offers an added design dimension that can be used to optimize the energy of the emitted photon.

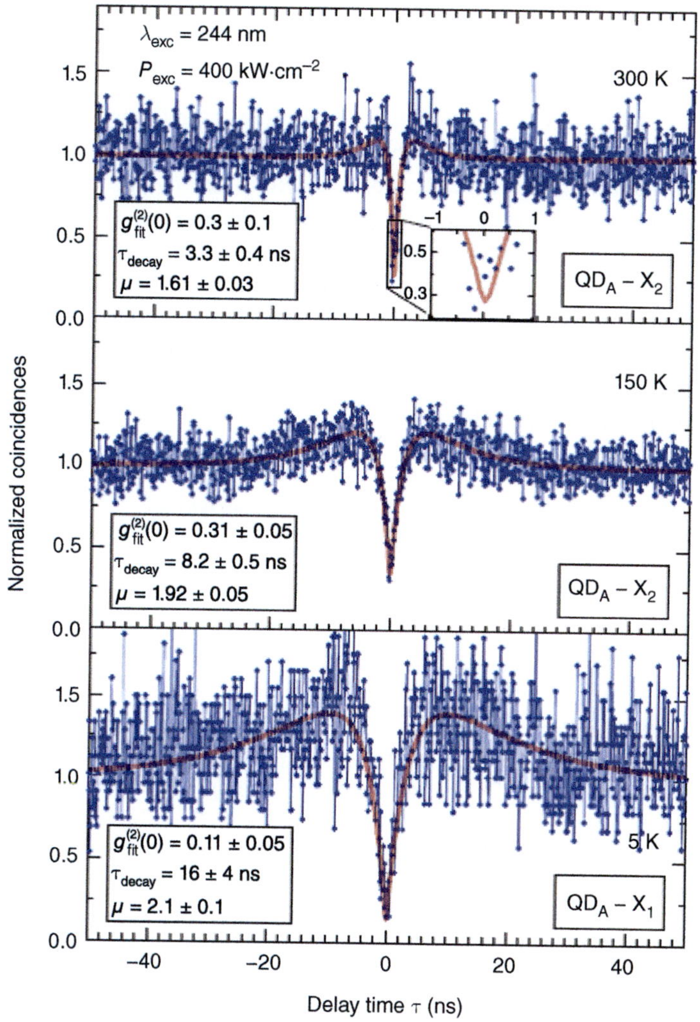

Figure 9.5 Temperature dependence of single-photon emission from GaN/AlN QDs. From Reference [34], used with permission.

It must be kept in mind that *a*-plane or *m*-plane growth is technologically more challenging compared to *c*-plane growth. So, while polarization engineering may be better for *a*-plane or *m*-plane growth, dislocation densities and stacking fault densities are higher for *a*-plane and *m*-plane growth compared to *c*-plane growth. Like any emitter, these dislocation densities have an adverse effect on emission in the quantum emitter also. This is because the dislocations lead to non-radiative recombination so that the radiative quantum emission is suppressed. Therefore, these trade-offs must be kept in mind while designing growth plane specific quantum emitters.

Spectral Diffusion in Quantum Dots

The energy of photons can suffer from spectral diffusion due to Coulomb interactions between the QD excitons and itinerant electrons in the surrounding medium. Optimization of Coulomb screening characteristics in the QDs is necessary to obtain pure, stable photon energies from group III-Nitride QDs.

Photon Linewidths

Coherent control of photonic qubits for quantum computing often requires spectral purity. The photon natural linewidth for group III-Nitride QD emission depends on the exciton lifetime. It can also be affected by phonon coupling with the excitons and can acquire a temperature dependence. Coupling with phonons also causes dephasing and reduced coherence, which can adversely impact quantum-computing fidelity. It becomes necessary to understand phonon dynamics in QD structures and determine how they vary from bulk phonon modes. It also becomes crucial to obtain the right coupling of nanostructure phonon modes with excitons. These aspects are currently the subject of intense research, and readers are encouraged to refer to the literature on this subject.

The polarization-induced QCSE also affects the photon linewidth. This is because the QCSE leads to separation between electron wave function and hole wave function in the QD. This, in turn, results in lower spontaneous emission amplitudes and lower linewidths.

Optically Pumped Versus Electrically Pumped Quantum Emitters

The QD giving quantum emission can be excited to its excited state optically or electrically. The optical pumping basically involves shining a light source like a laser on the QD until it gets excited with a hole in the valence band and an electron in the valence band. This kind of quantum emission is difficult to integrate in a big quantum system. The electrically pumped version of quantum emission works by electrically pumping electrons and holes into the QD from barrier layers. Electrons are pumped into the conduction band from an *n*-type barrier layer and holes are pumped into the valence band from a *p*-type barrier layer. This type of injection typically happens when the *p*-QD-*n* structure is forward biased.

The electrically pumped quantum emitters can be easily integrated in a quantum system and can lead to large-scale quantum computing. Heterojunction parameters like valence-band discontinuity and conduction-band discontinuity become crucial to obtaining the right level of electrical pumping at a given bias level so that the right kind of quantum emission can be obtained.

Photon Detection Properties

Since single-photon detection forms an integral part of a quantum-computing scheme that involves single-photon emitting QDs, integration using the same material system is of potential interest. Therefore, InGaN QD photon emitter systems can be integrated with InGaN single-photon avalanche photodiodes (SAPD) to obtain quantum-computing systems in the same material system. Since InGaN has a wide tunability of bandgap and effective mass, the impact ionization coefficients can be tuned over a wide range and good optimization can be attained for signal-to-noise ratio (SNR) of these single-photon detectors.

It must be kept in mind that InGaN alloy can have carrier localization effects for some In contents. This can have an adverse impact on photon detection because the photogenerated electron and hole may not be collected efficiently due to carrier localization. The effect of polarization must also be considered for single-photon detectors.

References

1 Holmes, M.J. et al. (2021). Pure single-photon emission from an InGaN/GaN quantum dot. *APL Mat.* 9: 061106.

2 Arita, M. et al. (2017). Ultraclean single photon emission from a GaN Quantum dot. *Nano Lett.* 17: 5.

3 Schweickert, L. et al. (2018). On-Demand Generation of background-free single photons from a solid-state source. *Appl. Phys. Lett.* 112: 093106.

4 Sun, X. et al. (2020). Single-photon emission from isolated monolayer islands of InGaN. *Light Sci. Appl.* 9.

5 Gao, K., Holmes, M., Arita, M., and Arakawa, Y. (2018). Measurement of the emission lifetime of a GaN interface fluctuation quantum dot by power dependent single photon dynamics. *Phys. Status Solidi.* 215 (9).

6 Zhang, L. et al. (2013). Single photon emission from site-controlled InGaN/GaN quantum dots. *Appl. Phys. Lett.* 103 (19).

7 Deshpande, S., Frost, T., Hazari, A., and Bhattacharya, P. (2014). Electrically pumped single-photon emission at room temperature from a single InGaN/GaN quantum dot. *Appl. Phys. Lett.* 105 (14).

8 Musial, A. et al. (2019). High-purity triggered single-photon emission from symmetric single InAs/InP quantum dots around the telecom c-band window. *Adv. Quantum Technol.* 3.

9 Le Roux, F. et al. (2017). Temperature dependence of the single photon emission from interface-fluctuation GaN quantum dots. *Sci. Rep.* 7.

10 Xia, S. et al. (2021). Enhanced single-photon emission from GaN quantum dots in bullseye structures. *ACS Photonics* 8 (6).

11 Sun, X. et al. (2020). Excitation and emission dynamics of a single photon emitting InGaN quantum dot in a photonic horn structure. *Superlattices Microstruct.* 145.

12 Holmes, M.J., Kako, S., Choi, K. et al. (2016). Single photons from a hot solid-state emitter at 350 K. *ACS Photon* 3 (4).

13 Cho, J. et al. (2018). Strongly coherent Single-photon emission from Site-controlled InGaN quantum dots embedded in GaN nanopyramids. *ACS Photonics* 5 (2).

14 Garevicc, Z. et al. (2017). Emission of linearly polarized single photons from quantum dots contained in nonpolar, semipolar, and polar sections of pencil-like InGaN/ GaN nanowires. *ACS Photonics* 4 (3).

15 Tamariz, S. et al. (2020). Toward bright and pure single photon emitters at 300 K based on GaN quantum dots on Silicon. *ACS Photonics* 7 (6).

16 Holmes, M., Kako, S., Choi, K. et al. (2015). Spectral diffusion and its influence on the emission linewidths of site-controlled GaN nanowire quantum dots. *Phys. Rev. B* 19: 115447.

17 Hanschke, L. et al. (2018). Quantum dot single photon sources with Ultra-low multi-photon probability. *NPJ Quantum Inf.* 4: 43.

18 Deng, J. et al. (2022). Disk-Shaped GaN quantum dots embedded in AlN nanowires for room-temperature single-photon emitters applicable to quantum information technology. *ACS Appl Nano Mat.* 5 (3).

19 Tribu, A. et al. (2008). A High-Temperature Single-Photon Source from Nanowire Quantum Dots. *Nano Lett.* 8: 12.

20 Sun, X. et al. (2019). Single-photon emission from a further confined InGaN/GaN quantum disc via reverse-reaction growth. *Quantum Eng.* 1 (3).

21 Springbett, H.P. et al. (2018). Improvement of single photon emission from InGaN QDs embedded in porous micropillars. *Appl. Phys. Lett.* 113

22 Tran, T.T. et al. (2016). Robust multicolor single photon emission from point defects in hexagonal boron nitride. *ACS Nano* 10: 7331–7338.

23 Jemsson, T., Machhadani, H., Holtz, P.-O., and Karlsson, K.F. (2015). Polarized single photon emission and photon bunching from an InGaN quantum dot on a GaN micropyramid. *Nanotechnology* 26 (6).

24 Arita, M., Le Roux, F., Holmes, M.J. et al. (2017). Ultraclean single photon emission from a GaN quantum dot. *Nano Lett. 17 (5).*

25 Wang, H. et al. (2019). On-Demand semiconductor source of Entangled photons which simultaneously has high fidelity, efficiency and indistinguishability. *Phys. Rev. Lett.* 122: 113602.

26 Holmes, M.J., Choi, K., Kako, S. et al. (2014). Room-temperature triggered single photon emission from a III-nitride site-controlled nanowire quantum dot. *Nano Lett.* 14 (2).

27 Wang, T. et al. (2017). Polarisation-controlled single photon emission at high temperatures from InGaN quantum dots. *Nanoscale* 9.

28 Miyazawa, T. et al. (2016). Single-photon emission at 1.5um from an InAs/InP quantum dot with highly suppressed multi-photon emission probabilities. *Appl. Phys. Lett.* 109: 132106.

29 Gao, K. et al. (2019). Spectral diffusion time scales in InGaN/GaN quantum dots. *Appl. Phys. Lett.* 114 (11).

30 Berhane, A.M. et al. (2017). Bright room-temperature single-photon emission from defects in gallium nitride. *Adv. Mater.* 29: 1605092.

31 Sun, X. et al. (2019). Single photon source based on an InGaN quantum dot in a site-controlled optical horn structure. *Appl. Phys. Lett.* 115: 022101.

32 Grosso, G. et al. (2017). Tunable and high-purity room temperature single-photon emission from atomic defects in hexagonal boron nitride. *Nat. Commun.* 8: 705.

33 Reid, B.P.L. et al. (2015). Non-polar InGaN quantum dot emission with crystal-axis oriented linear polarization. *Appl. Phys. Lett.* 106 (17).

34 Stachurski, J. et al. (2022). Single photon emission and recombination dynamics in self-assembled GaN/AlN quantum dots. *Light Sci. Appl.* 11.

10

Polaritons in Nitride Semiconductor Heterostructures

Strong Coupling between Excitons and Cavity Modes

The vast majority of light-emitting devices using group III nitrides are based on emission rates given by time-dependent perturbation theory of quantum mechanics. In these conventional devices, the emission rates are given by Fermi's golden rule of quantum mechanics. However, when the cavity quality factor (Q) is made very high, the coupling between electronic excitations, or excitons, and cavity photons enters the strong coupling regime. In this regime, the observable modes in the cavity are given by coupled modes. In other words, the exciton and the photon couple strongly to give two excitations called upper and lower polaritons, which are slightly displaced from both the exciton energy and the photon energy. The reader should keep in mind that the term polariton has also been used historically to also describe light propagation in materials with excitonic dispersion in the weak coupling regime. In this chapter, we are concerned only with the strong coupling regime, and therefore, only with strong coupling polaritons.

Conditions for Strong Coupling

The coupling between the exciton and the photon is given by the reduced Planck's constant times the Rabi frequency Ω (this is actually the angular Rabi frequency, but it is often called the Rabi frequency for simplicity).

$$g = \hbar\Omega \tag{10.1}$$

To gain insight into the physical meaning of the Rabi frequency, the reader is encouraged to read any good standard textbook on quantum mechanics that discusses light–matter interactions (or atom–photon interactions). The Rabi frequency is basically the matrix element of **A.p/m** between the ground state and the excited state divided by the reduced Planck constant. The condition for strong coupling is that the Rabi frequency must be much larger than the photon decay rate in the cavity.

$$\Omega \gg \gamma_{\text{ph}} \tag{10.2}$$

where γ is the cavity decoherence rate.

Group III-Nitride Semiconductor Optoelectronics, First Edition. C. Jayant Praharaj.

Another condition for strong coupling is that Ω should be much greater than the damping rate of the excited state Γ.

$$\Omega \gg \Gamma \tag{10.3}$$

When these two inequalities (conditions for strong coupling) are satisfied, the Rabi frequency is responsible for shifting eigenmode energies. If we use quantum states |0>|1 photon> and |1>|0 photons> as the basis in which to write the Hamiltonian, then the non-diagonal elements of the Hamiltonian matrix are proportional to $\hbar\Omega$. Here, |0> and |1> are the ground state and the excited state of the atomic system or the solid state system, respectively. This off-diagonal element basically shifts the energy of the eigenmodes to values different from the energy levels of the non-interacting system.

Energies of Polariton Modes

The energies of the polariton modes can be derived as (References [1–4])

$$E_{LP,UP}(k_{||}) = \frac{1}{2}\left[E_{exc} + E_{cav} \mp \sqrt{(2\hbar\Omega)^2 + (E_{exc} - E_{cav})^2}\right] \tag{10.4}$$

for quantum wells, where E_{exc} is the exciton energy and E_{cav} is the cavity photon energy. For quantum dots inside high-Q cavities, the corresponding equation is

$$E_{(LP,UP)} = \frac{1}{2}\left[E_{exc} + E_{cav} \pm \sqrt{(2\hbar\Omega)^2 + (E_{exc} - E_{cav})^2}\right] \tag{10.5}$$

These polariton energies are derived by diagonalizing a Hamiltonian matrix where the diagonal elements are E_{exc} and E_{cav} and the non-diagonal elements are $\hbar\Omega$. The reader is encouraged to do this simple derivation.

The problem is analogous to a system of coupled harmonic oscillators with non-zero coupling strength. The coupling leads to eigenmodes with frequencies split by the coupling strength.

Characterization of Polariton Modes

The energies of the polariton modes in a typical high-Q cavity can be observed by using photoluminescence (PL). The PL spectrum observed upon shining a laser light on this matter-cavity system shows the upper polariton branch (UPB) and lower polariton branch (LPB), as shown in Figure 10.1. It must be remembered that interactions with the lattice system (or phonons) lead to broadening of the polaritons in energy. Therefore, lower temperatures are better for observing the polariton branches using PL.

The splitting has angular dependence (angle measured with respect to the cavity walls) and can be observed in PL spectra. Before using this kind of matter-cavity system to build light-emitting devices like lasers, PL characterization gives a very good idea of what kind of frequencies or wavelengths can be obtained and the maximum temperature at which the polaritonic signatures can be obtained. The ability to use polaritons to obtain novel kinds of light emission depends on the

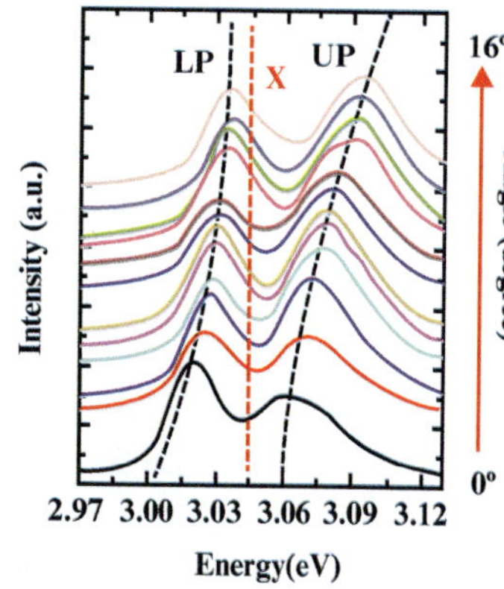

Figure 10.1 Typical polariton splitting in a high-Q cavity with indium gallium nitride (InGaN)/GaN quantum wells revealed by photoluminescence. From Reference [1], used with permission.

temperature at which these polariton branches are stable. As discussed later in this chapter (in the section on exciton binding energies), the binding energy of excitons is strongly material dependent, and only certain materials and certain device configurations lead to bound excitons at room temperature (RT). The E_{exc} relevant to this is the exciton energy (the electron-hole energy difference minus the exciton binding energy).

Polaritonic Lasing versus Photonic Lasing

Polaritonic lasing happens in the strong coupling regime when the polariton ensemble condenses to a quantum condensate. This is similar to the Bose–Einstein condensation seen for many bosonic gases. The reader can find descriptions of Bose–Einstein condensation in several excellent books on statistical mechanics. The light output in polaritonic lasing happens at a frequency corresponding to a polaritonic mode, as mentioned in the previous section. Polaritonic lasing differs from photonic lasing in that for photonic lasing, the lasing happens in the weak coupling regime and the laser light intensity builds up until the saturated current gain becomes equal to unity.

The reason why Bose–Einstein condensation becomes relevant to polariton gases (or polariton liquids) is that excitons are roughly bosonic in nature, while simple excited electrons in the conduction band are fermionic in nature. We will not prove the quasi-bosonic nature of excitons in this book, and the reader is referred to the vast literature on the topic of excitons for this. However, this bosonic nature of excitons means that multiple electron-hole pairs are allowed to exist in a specific excitonic state by the rules of quantum mechanics. This means condensates of exciton-polaritons are allowed by the rules of quantum mechanics.

One of the reasons for interest in polaritonic lasing is that it can potentially give intense, monochromatic light emission at low-threshold power or low-threshold current. The lasers discussed in Chapter 5 are all electrically pumped lasers, and the relevant quantity is the threshold current density. For polaritonic lasing, the reader must keep in mind that the technology is still developing, and a lot of the reports are highly exploratory. For this reason, many polaritonic lasing reports are based on optical pumping rather than on electrical pumping. For optically pumped lasers, the relevant threshold is the threshold power. Figures 10.2 to 10.4 show structures and characteristics of typical InGaN polariton lasers.

The physical reason for lower threshold current density or low-threshold power is related to the polariton density needed to obtain RT condensation of polaritons. The exact derivation of the

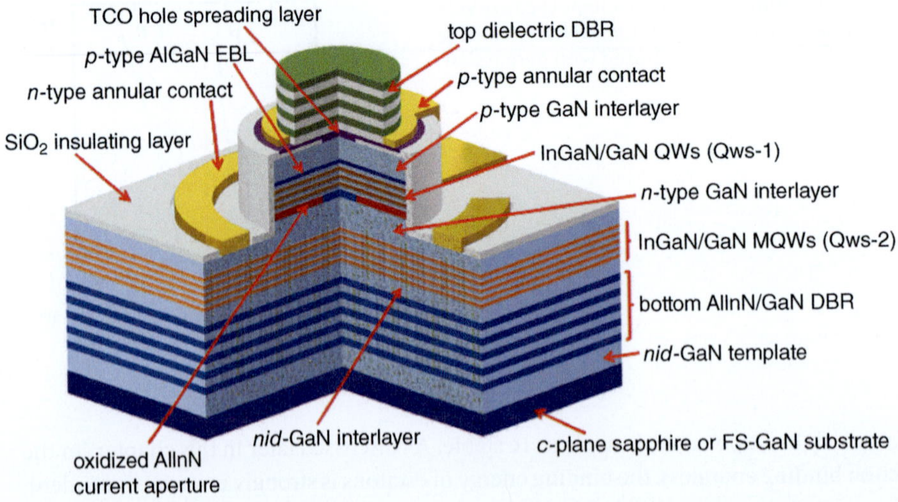

Figure 10.2 Typical device structure of electrically pumped InGaN-based polariton laser diode. From Reference [5], used with permission.

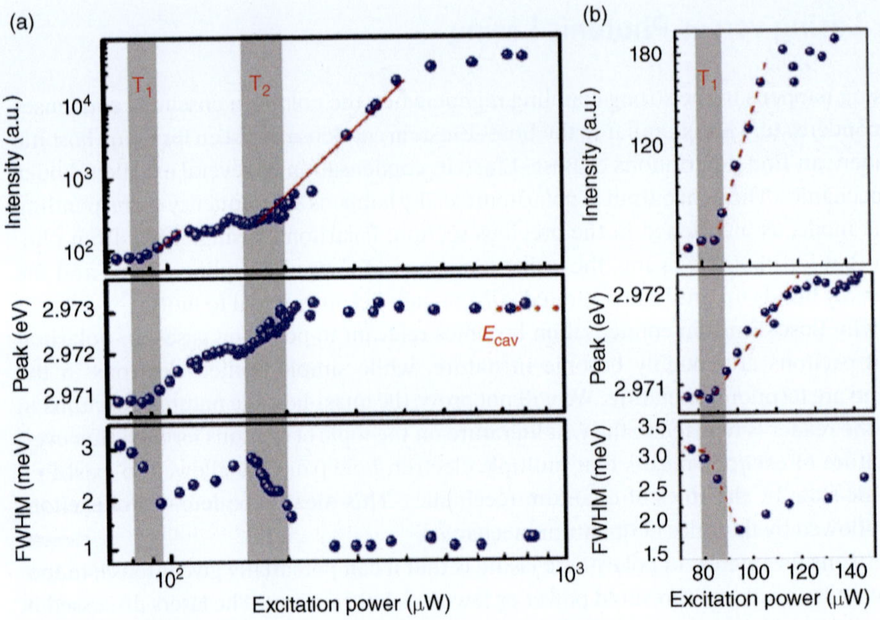

Figure 10.3 Polaritonic and photonic lasing thresholds. From Reference [2], used with permission.

polariton lasing threshold is mathematically involved and is still a subject of research. Polariton lasing can potentially offer threshold currents or threshold powers that are orders of magnitude smaller than that of photonic lasers. However, many reports in the literature see polariton lasing thresholds about half to an order of magnitude lower than in photonic lasing. This less drastic reduction in threshold current or threshold power can be attributed to excitonic inhomogenities in

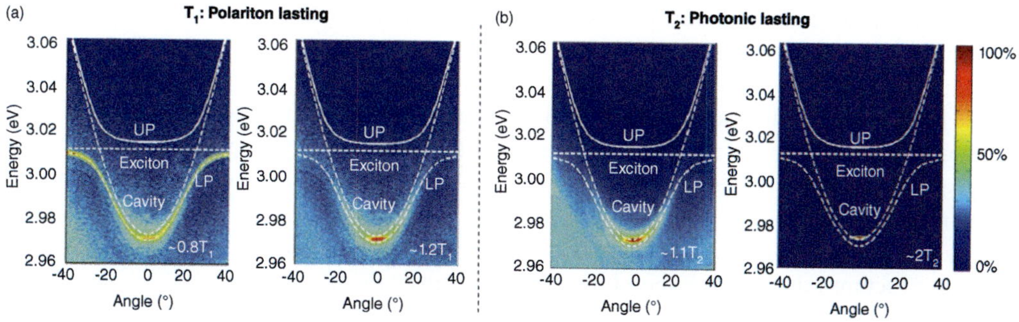

Figure 10.4 Intensity maps of polaritonic lasing and excitonic lasing. From Reference [2], used with permission.

group III-Nitride quantum wells. The reader can find details about polariton lasing reports in the literature in References [1–6]. Exotic polariton modes are described in Reference [7].

Exciton Binding Energies and Polaritonic Lasing

As mentioned before, the bosonic (actually near-bosonic) nature of excitons is crucial to achieving Bose–Einstein condensates of exciton-polaritons. The reader may be aware that exciton binding energies vary quite a bit from one semiconductor material system to another. The basic reason has to do with the fact that different semiconductors have different effective masses for electrons and holes as well as different dielectric constants. This leads to different Coulombic binding energies between conduction-band electrons and valence-band holes. The exciton binding energy in three-dimensional (3D; bulk) is given by

$$E_{\text{binding}}^{\text{exciton}-3D} = \mathrm{E}_{\text{Hartree}} \cdot \frac{\mu}{m_0}\left(\frac{\varepsilon_0}{\varepsilon}\right)^2 \tag{10.6}$$

where $E_{Hartree}$ is the Hartree energy (13.6 eV) from atomic physics, μ is the reduced effective mass of electron and hole, m_0 is the free electron mass, ε_0 is the permittivity of vacuum, and ε is the permittivity of the medium. This equation can be easily derived as the hydrogenic Coulombic binding between the electron and the hole.

To take an example, in the gallium arsneic (GaAs) material system, the binding energy of excitons is less than 25.6 meV, which is the thermal energy at RT. This means that excitons on GaAs are likely to be almost completely dissociated at RT. The physical mechanism by which this happens is phonon-exciton and phonon-electron interactions (or scattering). In GaN, the electron effective mass is about 0.25 m_0 (where m_0 is the free electron mass), compared to an electron effective mass of 0.067 m_0 is GaAs. The hole masses are also higher in GaN compared to GaAs. Further, dielectric constants are lower in GaN compared to GaAs. The combined effect of all this is that the exciton binding energy in bulk GaN exceeds 30 meV. This means these excitons are stable at RT.

Quantum confinement in quantum wells (QWs), quantum wires, and quantum dots (QDs) can further increase the exciton binding energy. The reader is referred to the literature for derivations of exciton binding energies in two-dimensional (2D), one-dimensional (1D), and zero-dimensional (0D; QD) systems. The increase in exciton binding energy with higher quantum confinement basically happens due to the modified Coulomb interaction between electrons and holes in these quantum-confined structures. This increase in exciton binding with quantum confinement becomes relevant when attempts are made to obtain polariton lasing in GaN/InGaN alloy quantum wells (with an InGaN QW layer) rather than aluminum (Al)GaN/GaN quantum wells. InGaN has lower exciton binding energy than GaN because effective masses are lower in InGaN than GaN. However, quantum confinement can make this binding energy higher than the thermal energy at RT. This is very helpful in obtaining RT polariton lasing using InGaN QWs, quantum wires, or QDs.

To gain quantitative insight into what was said in the previous paragraph, for 2D quantum-confined structures, the energy levels of the exciton are given by

$$E_n^{2D} = E_n^{3D} \frac{1}{\left(n + \frac{1}{2}\right)^2} \tag{10.7}$$

Using this equation we can see that quantum well excitons have a binding energy four times that of bulk excitons. Actually, Equation 10.7 is valid when the QW layer and the cladding layers (barrier layers) have the same dielectric constant as the 3D bulk material. This is often not the case. When the QW and barrier layers have different dielectric constants or permittivities, more careful calculations are required to solve for the exciton energy levels. Similarly, enhancements of exciton binding energies can also be calculated for 1D or quantum wire systems. Calculating exciton binding in QDs is even more involved, and it requires solving the effective mass Schrodinger equation along with Coulomb interaction terms.

One reason for preferring InGaN QWs or QDs over AlGaN QWs or QDs is that when we try to achieve electrically pumped polaritonic lasing, contacts to the barrier layers of the QWs or QDs become crucial. This is especially true for ohmic contacts to the *p*-type barrier layer. GaN/InGaN QWs and QDs are technologically much easier from this perspective compared to AlGaN/GaN QWs or QDs. Therefore, the fact that quantum confinement can lead to InGaN exciton binding energy of more than 26 meV is very significant for obtaining RT polariton laser technology in InGaN systems.

The technological significance of obtaining RT polariton lasing using InGaN QWs or QDs is that the visible range can be accessed, as opposed to mainly the ultraviolet (UV) range in GaN. The combination of a sufficiently large exciton binding energy and the accessibility of visible range means that ultra-low threshold polariton lasing can become relevant to many kinds of optical communication electronics.

There is also the added possibility of using ultrathin atomistic QWs, quantum wires, or QDs. While the growth technology for these can be challenging, it is feasible to obtain these atomistic quantum structures using current state-of-the-art molecular beam epitaxy (MBE) or atomic layer deposition (ALD) technology. Atomistic quantum-confined structures can lead to better confinement of electrons and holes as well as increased exciton binding energy.

Spontaneous and Piezoelectric Polarization Effects

Spontaneous and piezoelectric polarization affect the polariton spectrum by drastically shifting the exciton energy. The red shift typically caused in InGaN QWs by polarization-induced interface charges can lead to situations where the cavity design needs to be altered significantly to observe polaritonic light emission and polaritonic lasing.

QDs tend to have lower polarization-caused quantum-confined Stark-effect magnitudes. However, they are still significant enough to affect the polariton energy magnitudes and to effect the design of polaritonic lasers based on QD layers.

For polariton lasers built using QWs in high-Q cavities or multi-quantum wells (MQWs) in high-Q cavities, the use of non-polar *a*-plane InGaN or *m*-plane InGaN can eliminate polarization-induced red shifts. This can potentially help achieve low-threshold polaritonic lasing in some wavelengths. *a*-plane and *m*-plane InGaN growth technology is more difficult compared to *c*-plane growth technology. The dislocation density tends to be higher for *a*-plane and *m*-plane growth. This leads to non-radiative recombination, which can compete with radiative processes and adversely affect both polaritonic lasers and photonic lasers.

The coupling strength, or Rabi frequency, is also drastically modulated by the polarization-induced, built-in electric fields. The reduced overlap between electron and hole wavefunctions translates directly to reduced coupling strength. Therefore, the achievement of a strong coupling regime (SCR) is closely tied with the ability to minimize built-in electric fields and reduced electron-hole overlap. Therefore, *a*-plane and *m*-plane QWs and quantum wires can potentially give much larger coupling strengths and make it easier to obtain an SCR.

Optically Pumped versus Electrically Pumped Polariton Lasers

Both optically pumped and electrically pumped polariton lasing have been reported in group III nitrides in the literature. Electrically pumped polariton lasers can lead to better optoelectronic integration and to easier integration in communication systems. However, several technological challenges must be addressed to obtain low-threshold polariton lasing using electrical pumping. *p*-type and *n*-type ohmic contacts can prove a challenge in some group III-Nitride alloy composition ranges. Furthermore, valence-band discontinuities and conduction-band discontinuities must be in the right range to obtain the right density of exciton-polaritons at typical voltage bias ranges.

Electrically pumped quantum emitters can be easily integrated in a quantum system and lead to large-scale quantum computing. Heterojunction parameters like valence-band discontinuity and conduction-band discontinuity become crucial to obtaining the right level of electrical pumping at a given bias level so that the right kind of quantum emission can be obtained.

Electrically pumped polariton lasers can use vertical emitting configurations where the current injection from top is done by ohmic contacts and current spreading layers while the photon is emitted beyond the distributed Bragg reflector (DBR) layers used to obtain high-Q cavities.

Inhomogeneous Broadening in Polaritonic Lasing

In the case of InGaN QW-based polaritonic lasing, it must be kept in mind that InGaN QW widths have fluctuations, and there also are random alloy fluctuations that can affect the excitonic transition energy. These changes in excitonic energy due to fluctuations are often called inhomogeneous broadening in this context. The excitonic energy equations mentioned earlier in this chapter assume some fixed value of the effective mass of electrons and holes (and therefore, a fixed reduced mass). Alloy fluctuations mean that the fixed effective mass picture is not strictly valid, and therefore, the exciton energies can also have random fluctuations, which lead to inhomogenous broadening.

In the presence of inhomogeneous broadening, the coupling strengths or effective Rabi frequencies must be calculated based on some statistical distribution. There is evidence [8, 9] that increasing the number of InGaN QWs coupled with the cavity photon mode leads to an increase in the effective coupling strength (Rabi frequency). The role of inhomogeneous broadening on polariton lasing is still a topic of intense research, and the reader is referred to the broad literature on this topic.

Polariton Lasing in Quantum Heterostructure Nanocavities

Polariton lasing in quantum heterostructure nanocavities has been demonstrated using nanorod systems [5, 10]. These polariton lasers have smaller sizes, and they are potentially widely scalable. However, realizing electrically pumped versions of these nanocavity polariton lasers is an engineering challenge.

In InGaN/GaN-based or GaN/AlGaN-based nanocavity polariton lasers, the geometry of the nanorods as well as the effective spontaneous and piezoelectric polarization magnitudes will be crucial to the realization of low-threshold lasers.

References

1 Wu, J.Z., Shi, X.L., Long, H. et al. (2019). Large Rabi splitting in InGaN quantum well microcavity at room temperature. *Mater. Res. Express* 6 (7).

2 Wu, J., Long, H., Shi, X.L. et al. (2019). Polariton lasing in InGaN quantum wells at room temperature. *Opto-electron. Adv.* 2 (12).

3 Li, J., Chen, H., Yu, G. et al. (2023). Room-temperature exciton-polariton and photonic lasing in GaN/InGaN core-shell microrods. *Phys. Scr.* 98 (7).

4 Bhattacharya, P., Frost, T., Deshpande, S. et al. (2014). Room temperature electrically injected polariton laser. *Phys. Rev. Lett.* 112: 236802.

5 Won, S.J., Ko, M., Shim, H.S. et al. (2022). Polariton lasing of multiple-layered InGaN/GaN/AlGaN Axial heterostructure nanorods for tunable nanolasers. *ACS Photonics* 9 (10).

6 Iorsh, I., Glauser, M., Rossbach, G. et al. (2012). Generic picture of the emission properties of III-nitride polariton laser diodes: steady state and current modulation response. *Phys. Rev. B* 86.

7 Gong, S., Ko, S.M., Jang, M.H. et al. (2015). Giant rabi splitting of whispering gallery polaritons in GaN/InGaN core–shell wire. *Nano Lett.* 15 (7).

8 Glauser, M., Mounir, C., Rossbach, G. et al. (2014). InGaN/GaN quantum wells for polariton laser diodes: role of inhomogeneous broadening. *J. Appl. Phys.* 115.

9 Shi, X.L., Long, H., Wu, J.Z. et al. (2019). Theoretical optimization of inhomogeneous broadening in InGaN/GaN MQWs to polariton splitting at low temperature. *Superlattices Microstruct.* 128.

10 Kang, J., Song, B., Liu, W. et al. (2019). Room temperature polariton lasing in quantum heterostructure nanocavities. *Sci. Adv.* 5.

11

Plasmon-coupled Group III-Nitride Optoelectronic Devices

Coupling between Localized Surface Plasmons (LSPs) and Quantum Wells

The electric field of nanoparticles with localized surface plasmon resonances (LSPRs) has a distribution that leads to coupling with the excitations in quantum wells (QWs) when they are a small distance away. Coupling between QWs and LSPRs leads to an additional channel for radiating electromagnetic energy. Therefore, the internal quantum efficiency of devices like lasers and light-emitting diodes (LEDs) is modified and can experience large enhancements.

LEDs and Lasers Based on LSPR Coupling

Figures 11.1 and 11.2 show how photoluminescence intensity can be increased by adding LSP metallic nanoparticles (NPs) on top of indium gallium nitride (InGaN)/GaN multi-quantum well (MQW) nanorods (NRs). The enhancement of PL intensity due to LSP NPs also means that the emission intensity of LEDs at any bias level is increased. This means brighter LEDs can potentially be achieved using these NPs. References [2, 3] are further examples of NP-caused spectral shifts.

The internal quantum efficiency of MQWs coupled to plasmonic NPs can be written as

$$\eta_{int} = \frac{k_{rad} + k_{LSP}}{k_{rad} + k_{LSP} + k_{non-rad} + k_{abs-metal}} \tag{11.1}$$

where is the radiative recombination rate, k_{LSP} is the radiative recombination addition due to coupling with plasmonic NPs, $k_{non-rad}$ is the non-radiative recombination rate, and $k_{abs-metal}$ is absorption losses in the metallic NP. Reference [4] has details about ultraviolet (UV)-LED improvement due to NP-caused LSPR coupling.

PL is due to spontaneous emission processes. However, the enhancement of PL intensity with the addition of plasmonic nanoparticles is also an indication that stimulated emission in the QWs can be enhanced by adding plasmonic nanoparticles. This means that the gain spectrum of lasers made using these nanorods can potentially shift toward a higher gain at any given electrical bias. This can lead to lower threshold current densities for lasing and higher lasing power outputs.

Group III-Nitride Semiconductor Optoelectronics, First Edition. C. Jayant Praharaj.

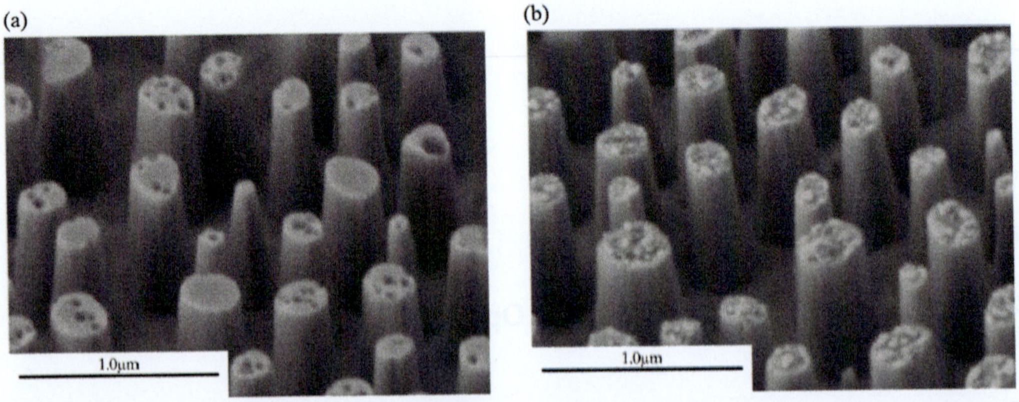

Figure 11.1 Bare nanorods (left) and nanorods with LSP nanoparticles (right). From Reference [1] (used with permission).

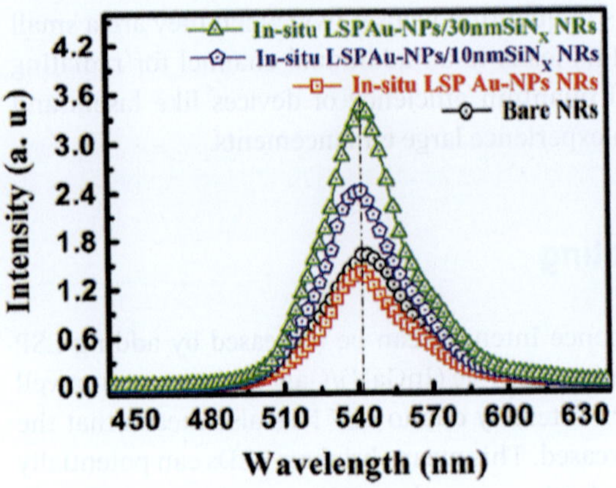

Figure 11.2 Photoluminescence (PL) spectra of various nanorod configurations with and without LSP nanoparticles. From Reference [1] (used with permission).

Biosensing Schemes Based on LSPR/QW Coupling

The modification of PL intensity of a QW system due to coupling with surface plasmon resonance (SPR) excitations in NPs provides a way to get high-sensitivity and integrable biosensors [5]. Typically, the QW is separated by an insulating barrier from the NPs.

Good sensing capability has been demonstrated for biomolecules like human immunoglobinG (hIgG) where binding of the biomolecule to NPs leads to shifts in PL intensities of the QW-NP system. High linearity with biomolecule concentration has been obtained. Also, the sensing scheme has been integrated into a microarray biochip configuration for highly versatile biosensing. Figure 11.3 shows some of the characteristics of this biosensor.

The well and barrier thicknesses should be optimized to obtain the best sensitivity in this kind of biosensing.

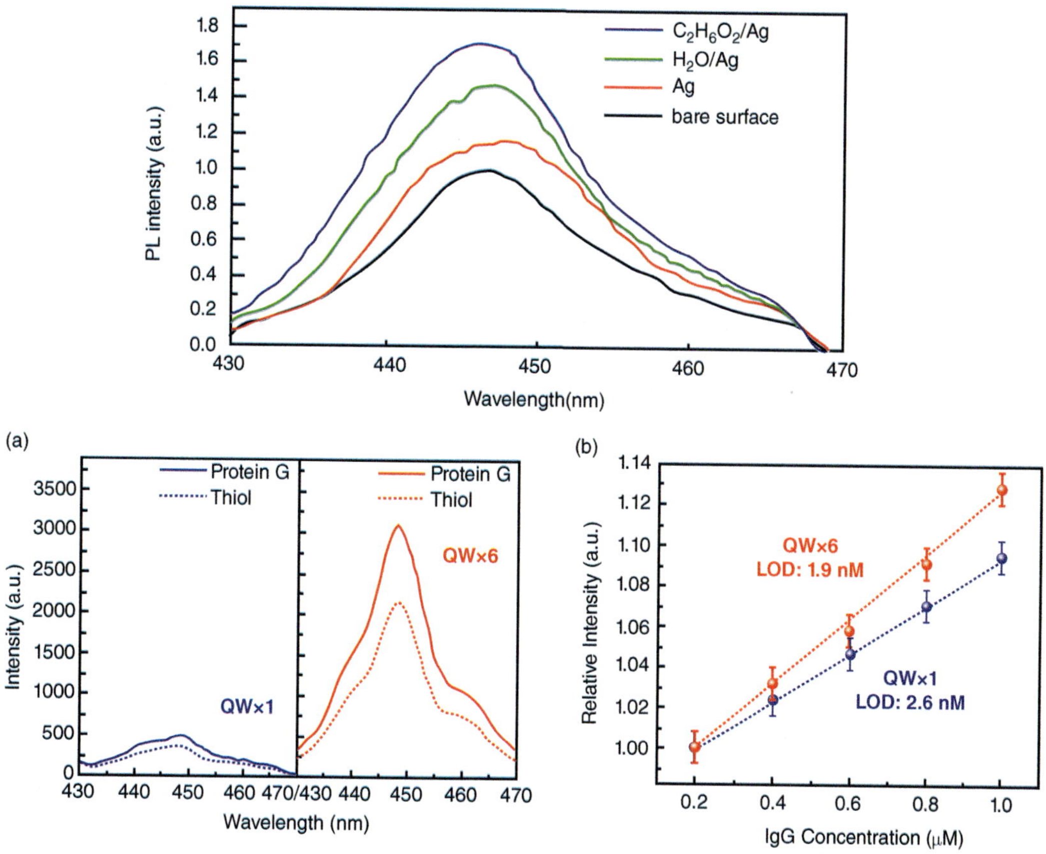

Figure 11.3 (Top) PL intensity modulation by surface conditions in QW-NP biosensor. (Bottom) Sensitivity to biomolecule hIgG. Reference [5], used with permission.

InGaN QW Substrates for Surface-enhanced Raman Scattering (SERS) Extended Hotspots

SERS is a technique where plasmonic particles can lead to giant enhancements of local electromagnetic (EM) field intensity (photon field intensity) so that the Raman signal (where a vibrational quantum of the molecule is emitted and a shifted-in-energy photon is emitted, while the original photon is absorbed) is amplified way above what can be expected in normal configurations. That means tiny concentrations of molecules, down to the single molecule limit, can potentially be sensed using this method. However, practical problems with making this method applicable for real-life situations exist.

One primary challenge for this technique is the fact that the hot spots (regions of enhanced EM intensity) are highly localized and this gives rise to lower sensitivity. The detection capability worsens because the incident laser radiation heats the molecules being detected so that they can migrate out of the regions of high intensity.

The use of InGaN QWs covered with NPs leads to extended hotspots so that even the tiniest concentration of molecules can be detected. This book does not derive the electromagnetic field profile for the QW NP-based SERS configuration. The reader should consult the references for details about simulation and experiments for improved SERS configurations with extended hotspots. Figure 11.4 shows typical SERS signal improvement by the use of MQWs.

The reader must be aware from Chapters 2 and 4 that InGaN QWs have built-in electric fields due to polarization. The number of wells in the MQW structure, the thickness of the wells, the thickness of the barriers, the composition of InGaN, and the spacing between the MQW and NPs are all design parameters that must be optimized to obtain the best extended hotspot profiles.

Similar hotspot size extension has also been achieved using InGaN quantum dot (QD) layers in proximity with NPs [7]. This method can catch single molecules with great reliability and can potentially extend the SERS techniques to new regimes of practical application.

InGaN Nanorods Plus Metal NPs for Water Splitting Using SPR Effects

Splitting water by photocatalysis is extremely useful to enable hydrogen fuel cell-based technologies for sustainable development. While a lot of semiconductor systems are used to split water by photocatalysis, the efficiency of hydrogen evolution from water tends to be low and is a challenge for technological applications.

Photocatalytic water splitting basically proceeds by the absorption of solar light at semiconductor nanorods (NRs), QDs, etc. by which electrons are excited into the conduction band and positive holes are left in the valence band. The electrons can combine with H^+ ions in the water to give H_2. The holes can combine with the OH^- ions in water to give O_2 and H_2. These reactions are photoelectrochemical reactions. Before these reactions can proceed, the electrons and holes must migrate from the bulk of the semiconductor to the surface.

InGaN is an attractive choice for photoelectrochemical water splitting because it has a tunable bandgap that can cover the entire solar spectrum. The band edge also has good alignment with the water redox potential and is thermodynamically suitable for water splitting. Typically, nanoscale structures like nanorods, nanowires, and QDs are preferable for water splitting because there is enough band bending to cause electrons and holes to separate. Separating electrons and holes is very useful because it prevents unwanted band-to-band recombination, which interferes with the photoelectrochemical process.

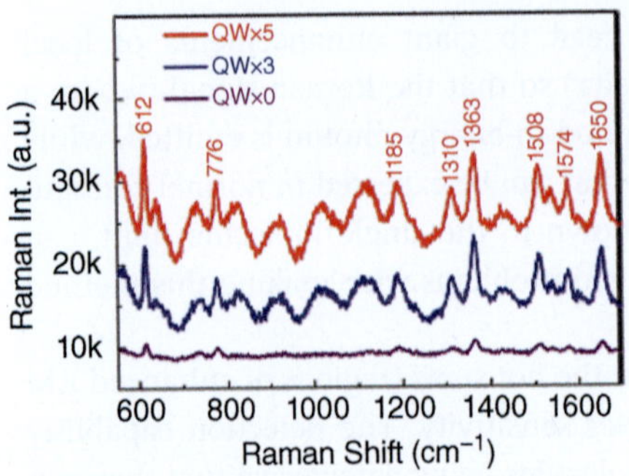

Figure 11.4 Increase in Raman signals with number of QWs in a QW-NP sensor. QWx5 has 5QWs, QWx3 has 3QWs and QWx0 has no QWs. From Reference [6], used with permission.

Further improvements can be obtained by functionalizing the InGaN surface with metallic NPs. These NPs have SPRs that couple with the semiconductor nanowire or QD. This coupling leads to enhanced absorption coefficients so that more of the solar energy or light energy is absorbed and more photogenerated electrons and holes are produced to cause water splitting. Some of the mechanisms involved in improving water splitting efficiency are hot electron injection enhancement caused by SPR, light-trapping enhancement, plasmon resonance energy transfer

(PRET), and light-concentration enhancement. These are all extensive topics by themselves, and the reader is referred to the literature on the subject for further details. Figure 11.5 shows band-energy alignments of silicon (Si) substrate, InGaN NRs, and plasmonic NPs. In hot electron injection enhancement, photons with energy below the InGaN bandgap but greater than the metal-InGaN Schottky barrier height can lead to photoelectrochemical water splitting. PRET is a plasmon-mediated energy transfer mechanism that leads to photogenerated carriers for water splitting. The light-concentration effect happens due to detailed electromagnetics of the NP-InGaN combination. Figure 11.5 shows the various energy transfer mechanisms at InGaN/NP interfaces.

Improved photocurrent responses are reported in Reference [8] when the InGaN NRs are decorated with NPs, as shown in Figure 11.6. It must be kept in mind that spontaneous and piezoelectric polarization can cause built-in electric fields, which can cause photogenerated carriers to flow along the rod axis instead of toward the NR surfaces. Proper NR design is necessary to obtain the best photoelectrochemical (PEC) efficiency using these schemes. The PEC efficiency is also affected by the exact size of the NPs, since the LSPRs and the LSPR-NR coupling depend on these dimensions. The optimal NP size for specific NR dimensions can be found experimentally.

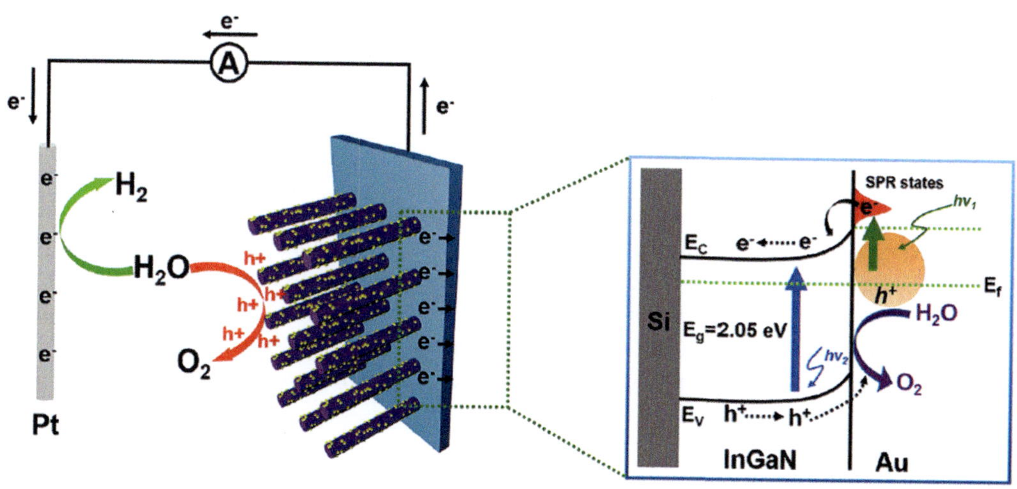

Figure 11.5 Energy transfer mechanisms at InGaN/NP interfaces. From Reference [8], used with permission.

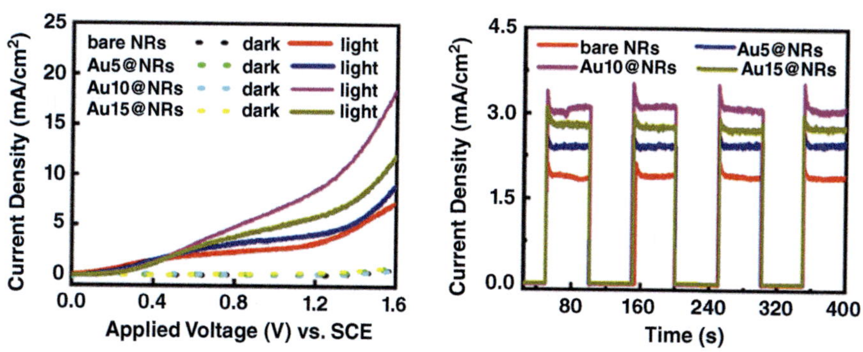

Figure 11.6 Photo-induced currents between InGaN/NP photoanodes and saturated calomel electrodes (SCEs). Several NP configurations show improvement over bare NR (no NP). From Reference [8], used with permission.

InGaN QDs Plus Metal NPs for Water Splitting Using SPR Effects

InGaN QDs offer another design space where they can be combined with metal NPs for water splitting using SPR effects. For this, QD-in-NR types of configuration can prove especially useful. For QD-based water splitting, the spontaneous and piezoelectric polarization effects also need to be kept in mind since they can affect where the photogenerated electrons and holes accumulate to cause the photoelectrochemical reaction.

References

1 Xing, J., Chen, Y., Liu, Y. et al. (2018). Enhancement of emission of InGaN/GaN multiple-quantum-well nanorods by coupling to Au-nanoparticle plasmons. *J. Appl. Phys.* 123.

2 Saleem, M.F., Peng, Y., Li, L. et al. (2022). Surface plasmon enhancement of an InGaN quantum well using nanoparticles made of different metals and their combinations. MDPI *Nanomaterials* 12.

3 Shugabaev, T., Gridchin, V.O., Komarov, S.D. et al. (2023). Photoluminescence redistribution of InGaN nanowires induced by plasmonic silver nanoparticles. MDPI *Nanomaterials* 13 (6).

4 Lee, J.W., Ha, G., Park, J. et al. (2020). AlGaN deep-ultraviolet light-emitting diodes with localized surface plasmon resonance by a high-density array of 40 nm Al nanoparticles. *ACS Appl. Mater. Interfaces* 12.

5 Chien, F., Lo, J.L., Zhang, X. et al. (2018). Nitride-based microarray biochips: a new route of plasmonic imaging. *ACS Appl. Mater. Interfaces* 10.

6 Chien, F.C., Zhang, T.F., Chen, C. et al. (2021). Nanostructured InGaN quantum wells as a surface-enhanced Raman scattering substrate with expanded hot spots. *ACS Appl. Nano Mater.* 4.

7 Zhao, F.Y., Nguyen, T.A.N., Tsai, C.W. et al. (2023). Catching single molecules with plasmonic InGaN quantum dots. *Adv. Opt. Mater.* 2300431.

8 Liu, Q., Shi, J., Xu, Z. et al. (2020). InGaN nanorods decorated with Au nanoparticles for enhanced water splitting based on surface plasmon resonance effects. MDPI *Nanomaterials* 10.

12

Photonic Integrated Circuits Using Group III-Nitride Semiconductors

Indium Gallium Nitride (InGaN)-based Monolithic Photonic Chips

There is a lot of interest in obtaining integrated functionalities with various optoelectronic/photonic components on the same chip for a wide variety of applications like biosensing, environmental monitoring, etc. InGaN is a material of choice for photonic integrated circuits (PICs) because it has tuneable bandgaps to suit a wide variety of applications, long device lifetimes, high physical and chemical stability due to rapidly maturing material growth, and device fabrication techniques.

A typical configuration can have a light-emitting diode (LED) or laser and a photodiode (PD) on the same chip [1–3], with the light from the LED (or laser) reflecting from an interface or surface into the PD. The amount of reflection can be modulated by a liquid, a solution, or a suspension. This achieves a complete sensor-on-a-chip PIC with a source, reflector, and modulator.

For some PICs with a laser source and detector on the same chip, it is preferable to have edge-emitting lasers of small dimensions. In this case, etched laser facets with smooth profiles (low roughness) need to be obtained. Reference [4] describes a process flow for obtaining AlGaN or InGaN laser diodes with etch facets. It uses a two-step dry and wet etching process. The dry etching involves inductively coupled plasma (ICP)-reactive ion etching (RIE) in chlorine. The wet etching uses tetramethylammonium hydroxide (TMAH). Figures 12.1 and 12.2 show the details of this method.

Other potential dry etching techniques for obtaining etched laser facets are reactive ion-beam etching, chemically assisted ion-beam etching, and atomic layer etching. This book is mainly on the device aspects, and we do not cover fabrication steps in great detail (although epitaxial growth methods are described in Chapter 3). The reader is referred to the vast literature on semiconductor processing and semiconductor manufacturing to obtain more detail about these various etching methods and their various advantages and disadvantages.

For PIC applications, cleaving the laser to obtain sharp profiles is not very convenient. When etched laser facets are used, there are strict requirements on the roughness and verticality of the etch profiles. The RIE+TMAH etching described earlier leads to smoother and more vertical facets. In Reference [4], the laser power output versus drive current curves show a clear increasing trend with higher TMAH etch time and get quite close to the power-current curve for the ideal cleaved-facet laser. Not only that, the threshold current also shows a clear decrease with TMAH etch time. This shows that current semiconductor fabrication technology is capable of achieving device dimensions and profiles suitable for PIC integration.

Group III-Nitride Semiconductor Optoelectronics, First Edition. C. Jayant Praharaj.

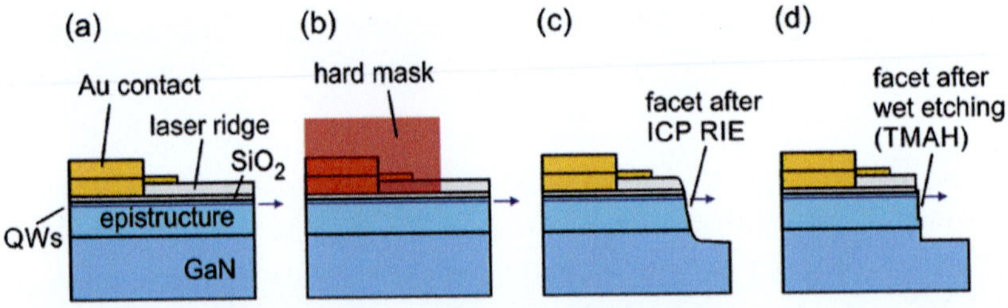

Figure 12.1 Dry plus wet etch to obtain etched laser facets for PICs. From Reference [4], used with permission.

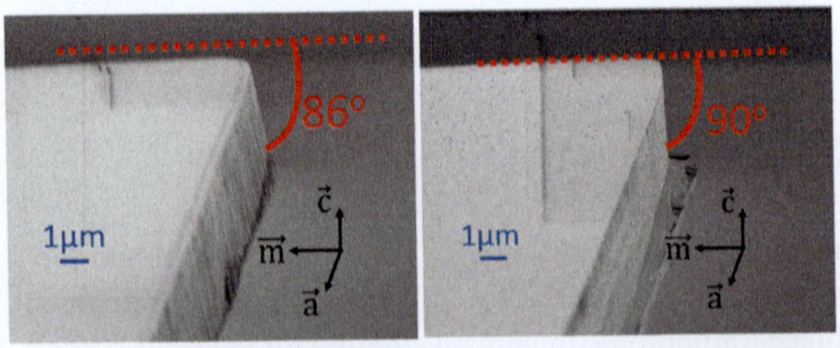

Figure 12.2 (Left) Laser facet after RIE etching. (Right) Laser facet after RIE and TMAH wet etching. From Reference [4], used with permission.

Similar considerations also apply to PDs in PICs. The etching methods mentioned above can be adapted to obtain ideal PD dimensions and profiles as well.

Photonic Integrated Circuits with Plasmonic Components

The plasmonic devices described in Chapter 11 can also potentially be realized in PIC configurations. The plasmonic nanoparticles (NPs) can increase the power output of lasers and LEDs in both vertical-emitting devices and edge-emitting devices. For PIC fabrication with plasmonic nanoparticles, the NP deposition or the metal patterning for obtaining NP must be compatible with the quantum well (QW) or multi-quantum well (MQW) parts of the device.

Obtaining PICs with plasmonic components in general, and plasmonically enhanced lasers and LEDs in particular, can lead to enhanced capabilities of sensors, environmental quality monitors, etc. The plasmonic enhancement of spontaneous emission and stimulated emission are particularly suitable to obtain laser-PD or LED-PD sensor configurations with ultra-high sensitivity.

Plasmonic sensors are highly versatile for biosensing and environmental sensing. Therefore, their integration in a PIC framework has a lot of technological benefits. It is still a matter of research which kinds of plasmonic sensors and plasmonic enhancers can be integrated in PICs using which kinds of semiconductor fabrication technologies. Also, the full scale of metallic NP dimensions and their morphologies are topics of research because the exact range of plasmonic enhancement and plasmonic sensing capabilities needs to be determined.

Exploring Modulators Using Nitrides for Easier Integration

Aside from sensing and quality monitoring, optical-based data processing and computation has been an aim of optoelectronic integrated circuits (OEICs) or PICs. For achieving logic gates optically in OEICs, modulators have to be used where the optoelectric effect is used to obtain constructive and destructive interference in Mach–Zehnder (MZ) interferometers. The ease of obtaining optoelectric modulation in interferometers depends on optoelectric coefficients and specialized materials like lithium niobate ($LiNbO_3$) are typically used.

To obtain interferometers on chips, one possibility is to deposit $LiNbO_3$ on GaN wafers and then pattern them lithographically. This is heterogenous integration, and it significantly increases the complexity of making the PIC. InGaN has a wide range of dielectric constants, and it can potentially offer frequency-dependent optoelectric coefficients that make achieving modulators, lasers, and PDs all on the same InGaN-based chip possible. More research is required to find if this is feasible using current device technology and current fabrication technology. For example, the behavior and fabrication of ultrathin MZ interferometer arms remain questions that need theoretical and experimental investigation.

While nitride-based modulators will immensely simplify the design and fabrication of OEICs in group III-Nitride technology, it is an open question if the modulator part can be achieved using group III nitrides. Therefore, the question of whether the nitride semiconductors can lead to more versatile and powerful OEICs compared to other semiconductors remains an open one. Reference [4] explores InGaN/GaN MQW directional couplers for visible light communication.

Combining Photonic and Electronic Components on the Same Chip

The implementation of system-on-chip using group III nitrides would be of great benefit if photonic components and electronic components can be combined on the same chip. Since digital circuits are easy to achieve in silicon (Si), one possibility is to use heterogenous integration with Si, where either the nitride layers are grown on a Si substrate or Si is deposited on sapphire/GaN. Heterogenous integration obviously comes with increased process complexity and the need for correct interfacing between the InGaN part and the Si part of the PIC.

The other way is to try to achieve an all-nitride integration with the electronic part of the circuitry, also built using InGaN or GaN layers. InGaN has a tuneable effective mass, tuneable bandgap, tuneable mobility, tuneable saturation velocity, etc so that we can potentially obtain versatile digital circuits by building the right kind of InGaN device structures. However, obtaining digital InGaN devices and circuits is not an easy task by any means. For example, if one wants to obtain field-effect transistors using InGaN, one must use oxide insulator deposition using a method like plasma-enhanced chemical vapor deposition (PECVD). One must keep in mind that the optical part of the chip is affected by this deposition step, and care must be taken to ensure that the deposited oxide is removed from the optical part of the chip once the field-effect transistor insulator has been achieved in the electronic part of the chip.

Monolithically Integrated Multi-color LED Display on a Single Chip

Nanowire InGaN LEDs grown on Si substrate has been used in Reference [5] to obtain a multi-colored LED display on a single chip. In this study, selective area growth (SAG) has been used to obtain a range of nanowire diameters with InGaN LED layers embedded in the nanowires. Indium incorporation depends on the diameter in this SAG-based process so that the LEDs with different diameters have peak emissions at different wavelengths, as shown in Figure 12.3.

Excellent robustness against mechanical stress has been demonstrated for these multi-colored LED chips in Reference [5]. This is important because there can be residual stresses from process variability as well as thermal stresses due to operating conditions. The ability to obtain stable multi-colored LED chips is a definite advantage for a wide range of applications.

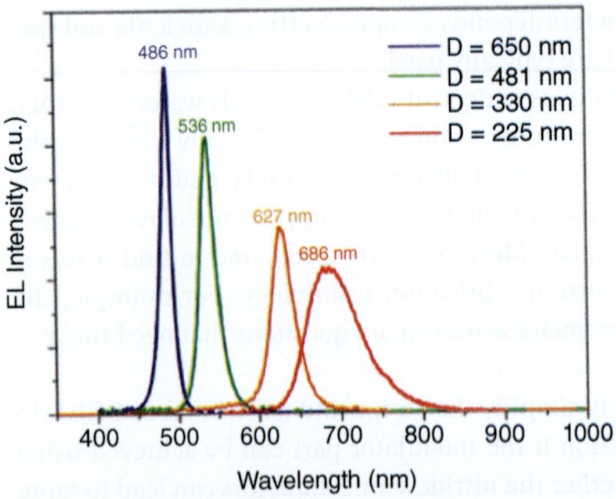

Figure 12.3 InGaN nanowire diameter-dependent electroluminescence (EL) spectra. From Reference [5], used with permission.

Spontaneous and Piezoelectric Polarization Effect on Multi-color LED Chips

It must be kept in mind that spontaneous and piezoelectric polarization are very relevant to the spectra obtained in this monolithically integrated multi-color LED chip. The strain relaxation can be very different in SAG growth as opposed to large area growth. Therefore, the exact state of strain depends on the growth mechanism as well as the InGaN layer composition and the layer thicknesses. This means that the spontaneous and piezoelectric polarization in these LEDs may need to be understood empirically to arrive at the best designs.

Also, it must be kept in mind that the spontaneous and piezoelectric polarization can show local fluctuations due to alloy fluctuations in the InGaN layers. The alloy fluctuation magnitudes and details can be different for different diameters of the nanowires. Theoretical understanding of these polarization fluctuations is necessary to understand the effect on quantum states and on emission spectra.

References

1 Chen, L., An, X., Jing, J. et al. (2020). Ultracompact chip-scale refractometer based on an InGaN-based monolithic photonic chip. *ACS Appl. Mater. Interfaces* 12.

2 Chen, L., Wu, Y.P. and Li, K.H. (2020). Monolithic InGaN/GaN photonic chips for heart pulse monitoring. *Opt. Lett.* 45 (18).

3 Gibasiewicz, K., Kafar, A., Schiavon, D. et al. (2023). InGaN laser diodes with etched facets for photonic integrated circuit applications. MDPI *Micromachines* 14.

4 Xin, L., Yun, L., Xu, W. et al. (2022). Silicon-based InGaN/GaN multiple quantum well waveguide directional coupler photonic integrated chip for visible light communication. *J. Electron. Inf. Technol.* 44 (8).

5 Qu, J., Wang, R., Pan, P. et al. (2023). Optically and electrically invariant multi-color single InGaN/GaN nanowire light-emitting diodes on a silicon substrate under mechanical compression. *Nanoscale* 15.

13

Conclusion

The field of group III-Nitride optoelectronic devices has emerged from intense research activity over the last 30 years. However, many material and device aspects are still the subject of intense research. The growth of template-based substrates is a subject of research as focus shifts to better the optimization of light emitters, photodetectors, and photovoltaics. Distributed Bragg reflector lasers for achieving high spectral purity in lasing is a subject of intense activity and a lot of progress has been made. The ability to get combined red, green, and blue emissions using hybrid light-emitting diode (LED) structures is also a topic that gets a lot of attention. The optimization of quantum-dot growth has a lot of scope for improvement as both layer composition and dimensions need to be optimized for achieving emission in the desired wavelength range with maximum efficiency.

The subject of dislocations also has a lot of scope for improvement. Threading dislocations affect light emission and quantum efficiencies badly. While Matthews–Blakeslee-type limits provide a guide as to when to expect dislocation formations and how thin epitaxial layers should be to have low dislocation formation, these kinds of models are thermal equilibrium models. However, a lot of epitaxial growth happens in the regime of non-equilibrium kinetics. Therefore, the subject of critical thickness and strain engineering is still very empirical. There is a need to come up with non-equilibrium growth models to compare with experiments and technological progress.

Substrate engineering is still a challenging field as the device epilayer quality depends on it. A lot of progress has been made in the electronics world in this area, and some of that progress spills over into the optoelectronics world as well. Growth along non-polar and semi-polar axes needs more maturing; it also needs to extend from the indium gallium nitride (InGaN) area more into gallium nitride (GaN) and aluminum gallium nitride (AlGaN). Substrate engineering and epitaxy must be understood together and in a better way in this context.

The reliability of LEDs and lasers is a topic that can see tremendous advances in coming years. Topics like the coalescence of dislocations and vacancies need to mature further to provide insights into light emission reliability.

Polarization engineering in photovoltaics is an area that needs better theoretical and experimental understanding. The development of intermediate band photovoltaic cells needs more development in the context of InGaN. For example, the use of quantum layers needs to be understood better in terms of the trade-offs between absorption characteristics and recombination.

The subject of optoelectronic integrated circuits is also an area that less attention has been given to compared to discrete LEDs, discrete lasers, and discrete photodetectors. Spontaneous and piezoelectric polarization offer the scope for engineering emission, absorption, and modulation

Group III-Nitride Semiconductor Optoelectronics, First Edition. C. Jayant Praharaj.

characteristics to achieve optoelectronic based computation. Also, quantum layers can potentially be used for achieving engineered non-linear optical characteristics and enhanced integrated optoelectronics (OEIC) design.

Quantum emitters, for example, single-photon emission from quantum dots has a lot of scope for improvement in terms of spectral uniformity and good coherence characteristics. Quantum emitters are of interest in both the InGaN band as well as the GaN and AlGaN band of energies. Therefore, quantum-dot growth of the latter kind of layers is of great potential interest in quantum technologies. The optimization of quantum-dot emitters in terms of wavelength, spectral purity, and coherence is likely to dominate this field in the years to come.

Index

Printed and bound by CPI Group (UK) Ltd, Croydon, CR0 4YY
17/09/2026
14964314-0007